# Amini Mudumbi Jacques

## Computer Aided Drafting (CAD)

Amini Mudumbi Jacques

# Computer Aided Drafting (CAD)

## Module and Practical Work

ScienciaScripts

**Imprint**

Any brand names and product names mentioned in this book are subject to trademark, brand or patent protection and are trademarks or registered trademarks of their respective holders. The use of brand names, product names, common names, trade names, product descriptions etc. even without a particular marking in this work is in no way to be construed to mean that such names may be regarded as unrestricted in respect of trademark and brand protection legislation and could thus be used by anyone.

Cover image: www.ingimage.com

This book is a translation from the original published under ISBN 978-620-6-70212-2.

Publisher:
Sciencia Scripts
is a trademark of
Dodo Books Indian Ocean Ltd. and OmniScriptum S.R.L publishing group

120 High Road, East Finchley, London, N2 9ED, United Kingdom
Str. Armeneasca 28/1, office 1, Chisinau MD-2012, Republic of Moldova, Europe
Printed at: see last page
ISBN: 978-620-7-14037-4

# Liminaires

## 1. Course viewpoint and rationale

*You use stone,*

*wood and concrete, and with these materials,*

*you build houses and palaces.*

*It's construction. Ingenuity at work.*

*But suddenly you touch my heart, you make me feel good,*

*I'm happy and I say: that's beautiful. This is architecture.*

*Art coming in.*

**(*Le Corbusier*)**

The introduction of **Computer-Aided Design** underlines the fact that today's man has not fundamentally changed the way he communicates. Even though he invented the alphabet, the printing press and the computer long ago, he is still a talented draughtsman. Children first express themselves through drawings. Much later, it is still through drawings that an architect imposes his art, that a mechanic brings new machines into being. We've come to understand the importance of this means of communication because it's a universal, visual language... more spontaneous perhaps! Don't we often hear people say "a drawing is better than a long speech" to shorten an explanation, or on the contrary, "do you need a drawing?" to those who don't understand?

With the invention of Computer Aided Drafting (CAD) software, it soon became apparent that certain building design tasks could be carried out more quickly and clearly using computer tools. Large architectural firms were quick to equip themselves, using "computers mainly as huge databases to optimize the efficiency of producing drawings with many repetitive elements". The use of this tool quickly spread to architects. CAD tools were used to produce project production documents (plans, sections, elevations, etc.). These changes in practice spread progressively from the late 60s for the forerunners to the late 90s, when the vast majority of architectural practices were equipped with computer-aided drafting tools.

In fact, many fields, as diverse as economics, the arts, ... architecture, benefit from this tool. The versatility of the computer means that architects can integrate it into their project design at totally different levels. To put it mildly, the applications are very diverse: project representation, the development of new construction methods, the calculation of complex structures, improved communication between the various parties involved in the project, increasingly automated management of different tasks... but it is only more recently that computers have been involved in the design and manufacturing process.

Since the computer was already present in our midst, **Negroponte** imagined transforming it into an alter ego whose differences in intelligence and logic could enrich the dialogue throughout the architectural process in a certain human conviviality. Thus, among the most frequent computer manipulations, that of "undo" has extremely interesting implications for computer architecture. It allows you to return to a previous state after a manipulation error.

By integrating this conceptualization stage with the computer processes involved in shaping these concepts, we can clearly say that **Computer Aided Drafting** (CAD) is a discipline that enables technical drawings to be produced using computer software. It differs from image synthesis in that it does not involve the rendering of a digital model, but the execution of graphic commands (lines, various shapes, etc.). As a result, in CAD, the mouse and keyboard replace the pencil and other drawing instruments. From all this, from what we can understand of objectivity, a device or the drawings produced are most often produced in vector mode (coherent strokes), whereas the computer-generated image is an association of independent bitmap pixels. In other words, CAD software assigns coordinates (X, Y for 2D plans and X, Y, Z for 3D models). Each element of a drawing is referred to as an entity, and each entity therefore contains properties such as color, thickness, layer, line type and so on.

It's true that CAD is first and foremost a tool for information technology, essentially a practical tool for document management, facilitating editing, archiving, reproduction, data transfer and so on. Although there are as many CAD programs as there are trades that use drafting. Today's mechanics, architects, electricians and surveyors all have access to tools that facilitate the creation of drawings and schematics, with business-oriented commands and adapted databases, as well as component catalogs supplied by manufacturers.

So there's a relationship between CAD and the digital technique of DMA (Digital Matte Artist).

On the whole, CAD is often confused with CAO (Computer-Aided Design): the primary function of CAO is not to edit drawings. It's a computer tool, often linked to a trade, operating in an object-oriented language, and enabling the virtual organization of technical functions. This then enables simulation of the behavior of the designed object, with any editing of a drawing or schematic being automatic and incidental.

It is now well established that computer technology is an indispensable element in the architect's current and future practice. In both project design and communication, the use of computers brings precision and time savings. However, to create a drawing quickly (economy is essential) and efficiently (a mistake can be fatal), it is important to master a certain number of tools. The finely-sharpened pencil is one such tool, and remains so in certain phases of a project. But the computer has now come a long way.

The ease with which it can be modified, the rigor and precision with which it can be traced, and the ability to animate the image, make the computer an essential tool.

Numerous drawing and image-generation software programs have already been developed. As a result, most schools of architecture and universities, as well as those training future technicians, have included a Computer Aided Drafting (CAD) course in their curriculum. The words speak for themselves: the course teaches the use of computer-aided drawing tools.

Finally, CAD refers to the activity of drawing up plans, which, in the past and still today in design offices where old, non-digitized "projects" are reworked, was done on a drawing board, using a ruler, tee, square, pencil and compass. This was traditional drawing, or instrument drawing, or technical drawing. In CAD, hardware and computer programs have replaced the usual and historic tools, but the objective and result are identical.

The aim of this course is to deepen the knowledge acquired in the field of building design, using computer tools, and to design using software such as Archicad, and even to carry out practical operations such as: opening-closing; adjusting the working environment, tools and palettes; wall-post-slab-metal-roofing-photographic rendering.

## 2. Pedagogical approach

The course is based on a process of updating the theory, practice and knowledge of the prerequisites for using a computer to carry out routine tasks, as well as notions of building design. On the one hand, this course provides an overview of this field of study, with an understanding of the implications of software for the creation of architectural drawings. In addition, it explores new features such as opening-closing, work environment settings, tools and palettes, and even wall-post-slab-roofing-photo-rendering, which currently characterize architectural practices for various house designs. All in all, the course favors an interactive approach.

## 3. Prerequisites

Know how to use a computer for everyday tasks...

## 4. Teaching aids

The student receives a booklet written by us.

## 5. Evaluation mode

The final exam is double-sided, i.e. theoretical and practical, and aims to verify the student's mastery of the concepts and notions studied in the course. The questions will be specifically technical and oral, requiring a very good knowledge of the tricks developed in the auditorium. It is made up of six questions, each of which will be marked out of 3, 2 or 4 points, with the total number of points for the exam then being 20, and 5 points for the student who has attended ¾ of the course sessions, 15 points for practical work and questioning to obtain an overall mark out of 40.

**Contents**

Open-close ;

Working environment setting ;

Tools and palettes ;

Wall-Post-Slab-Masonry-Roofing-Picture rendering ;


**Selected bibliography**

[1] Barbey, G., (1971), L'enseignement assisté par ordinateur, Collection E3, Casterman, Tournai, belgique ;

[2] Bignon, Jean-Claude, Jean-Claude Halin, and Sylvain Kubicki. Conception architecturale numérique et approches environnementales: Actes du 3e séminaire de conception architecturale numérique. Nancy: Presses Universitaires de Nancy, 2009.

[3] Bourdonnais, Sébastien. "Technological sensibilities: experimentation and exploration in digital architecture 1987-2010."

[4] Farel, Alain, and Denis ALKAN. Sur l'informatisation de la phase d'esquisse en architecture. Paris: BRA, 1987.

[5] "Impact of a virtual sketching environment and an early 3D model on architectural design activity." Revue d'Interaction Homme-Machine 8, no. 2 (2007): 65-98

[6] "Computers and architecture." Le Moniteur Architecture cahier N°1, no. 18 (March 1991): 75-94.

[7] CAD in architecture. Traité des nouvelles technologies, Hermès, Paris, p.59 ;

[8] Negroponte Nicholas. The Architecture Machine, towards a more human environment, MIT Press, Cambrigde and London, (1970) 1972 ;

[9] SPERBER Karl-Heinz; 2007. ArchiCAD 10. ...

[10] WILK, Eric. 2005. ArchiCAD: From Getting Started to Plotter;

[11] WILK, Eric. 2006. ArchiCAD: FROM CAD TO METER. ...

# INTRODUCTION

ArchiCAD is a modeling application that enables architects to design buildings productively based on the Virtual Building™ (Digital Building) concept. With ArchiCAD, architects can concentrate on their drawings whether working alone or as part of a team using the Teamwork function, exchanging data with consultants and specialists from other professions.

You can install ArchiCAD 19 whether you need to work with ArchiCAD as a single user or as part of a shared project.

Computer-aided drafting (**CAD) is the** discipline of producing technical drawings using computer software.

ArchiCAD 19 is a complete and flexible BIM solution for the architectural design market. BIM (*Building Information Modeling) is* the process of creating and managing building data throughout its lifecycle. In general, it uses three dimensions in real time, enhanced by dynamic modeling software to increase productivity in architectural design and construction. The process produces the **Building Information Model** (also abbreviated **BIM**), which encompasses the construction geometry, spatial relationships, geographic information, quantities and properties of component buildings.

Every architectural project contains multiple models and hundreds of drawings. Creating and managing this mass of information is a major and decisive task, and the automation of procedures is essential. As for mastering it, it's mainly a question of permanent, fluid navigation between multiple documents and information (graphic or geometric elements, indexes, quantities, links with other software...), obviously updated and synchronized in real time.

From the simplest 2D detail to the latest cross-section, facade, perspective or project overview, ArchiCAD 19 offers an efficient management system. Freely or automatically structuring the list of all available elements, ArchiCAD 19 gives instant access to everything, and even offers archiving, sharing or broadcasting, whether immediate or scheduled.

Let's just say that drawing software has evolved from 2D drawing to 2D/3D automation. Archicad was developed on this principle, taking into account the very basis of an architect's work, which is none other than the back-and-forth between the various technical components of an architectural project (plans, sections, facades), because in no case are we advocating work concentrated on the plan, for example, from which the facade and volumetry would only be results; Archicad, with its ability to simultaneously have 3D at hand, facilitates architectural reflection, develops creativity and provides visual feedback on its work. Archicad's development is based on 3D building simulation, better known as MIB (building information modeling). Archicad is now compatible with Microsoft Windows and Mac OS monitors. It allows you to model a building in 3D and to place various documents necessary for its construction (plans, sections, elevations, perspectives, object lists, etc.) on points. Its creation is attributed to **GABOR BOJAR**, who released the very first version in 1982, this software was written using the C language.

In both 3D and 2D, ArchiCAD's concepts enable you to design, assemble and represent, in a single model, the synthesis of project components, simulate their interactions, quantify them, and view the whole to understand the details. All entries can be made in plan, façade, cross-section or 3D view, with ArchiCAD providing full interactivity in real time.

The software can read the following file formats, among others: (DWG, DXF, IFC, CINEMA 4D (d), Rhino 3D Model (d), 3D Mesh file format (en), Keyhole Markup Language, Keyhole Markup Language Zipped (d), Sketchup (d), DAE, STL, Portable Document Format, AutoCAD Design Web, Fomat (d), Windows Bitmap, SDSC Immage Tool Run-Length Encode Bitmap (d), JPEG File Interchange Format, JP2 (d), GIF, Tagged Image File Format, Portable NetWork Graphics, PICT and ICO etc.

# INSTRUCTIONS FOR USE

This booklet is made up of 4 practical sheets that will help you bring your architectural projects to life. The first three sheets explain how to open and close ArchiCAD 19, how to set up the working environment and how to use the tools and palettes, with methodological, pedagogical and practical tips, while the last sheet explains how to design a building drawing, using wall-post-slab-masonry-roofing-photo-rendering, with detailed tips and even screenshots **to** illustrate these tips and make the course easier to follow. To help you find your way around, keep in mind that each case study is divided into the following sections:

- ☑ Each **practical data sheet** is introduced by a **Headline** illustrating a number of case studies, as well as all the elements developed to be as close as possible to practical characteristics;
- ☑ **Tips** to help you follow all the steps, even without a guide or even alone at home;
- ☑ Each practical data sheet groups together various concepts covered in the form of **remarks** and/or **Good to know,** as well as certain key words developed as **Turnkey**; but you will also find, when necessary, the general operation enabling you to create other projects;
- ☑ Each practice sheet is developed with **screenshots** taken as **case studies** to illustrate everything that has been titled. These screenshots can be used as they are, provided they fit in with your student approach and meet the objectives you've set for the next day's work. If this is not the case, they may inspire you to adapt them...
- ☑ **It's your turn to create**: whatever the objective of your student approach to the development of your various building drawings, we encourage you to adapt these existing teaching tips so that they correspond to future reality as well as

to each theme tackled. With this in mind, these methodical tips will give you the keys to adapting or creating the type of architectural drawings proposed.

**Please note**: although they are presented in various practical sheets, these practical tips are not specific to the example to which they are attached here: they are reusable (and should be reused for other cases too, to get you to create your various drawings as well as develop other elements needed for your buildings!) in all circumstances!

## INSTALLATION OF ARCHICAD 19

After inserting the disc, flash or CD/DVD, a presentation showing ArchiCAD's potential is automatically launched. If this presentation does not launch automatically on Windows, double-click on the **"Setup"** application in the root directory of your disc, flash or CD/DVD.

**Case study :**

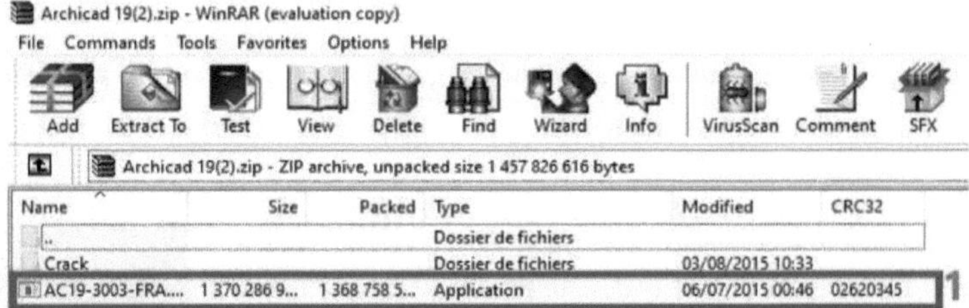


*Figure 0.0: ArchiCAD 19 software*

On startup, a window appears, asking you to select the location for your project. To do this, click on **Browse** to choose the location in which your project will be placed, then click on **Extract**.

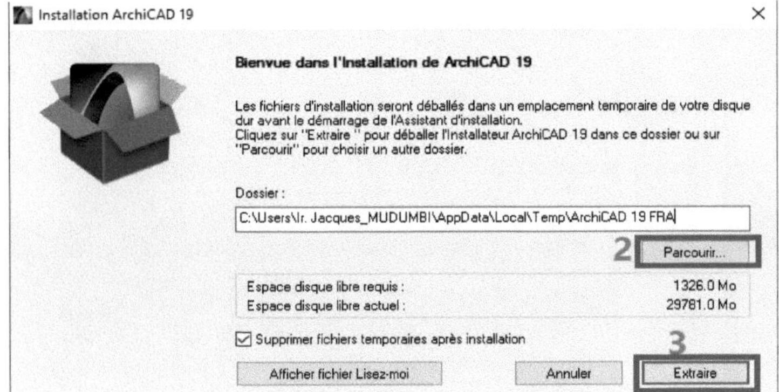

*Figure 0.1: ArchiCAD 19 installation*

Wait until the application finishes installing the **Archicad 19 installer package**.

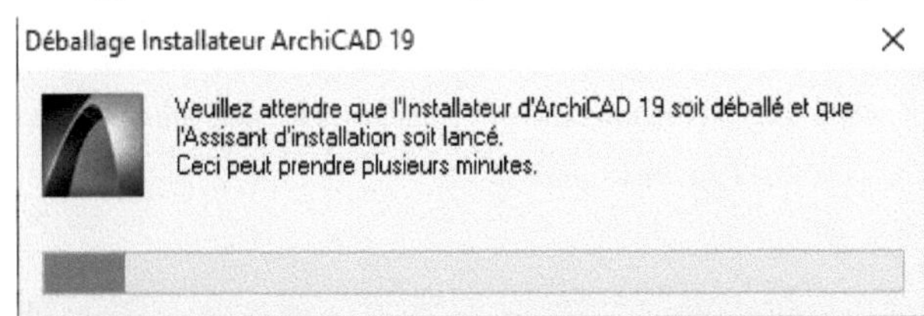

*Figure 0.2: Unpacking ArchiCAD 19 installer*

**Under Windows**: Click on **Install GRAPHISOFT ArchiCAD 19** in the menu to launch the ArchiCAD installation wizard, which first checks your Java environment. The ArchiCAD installer will automatically install Java 8 if it is not present on your computer.

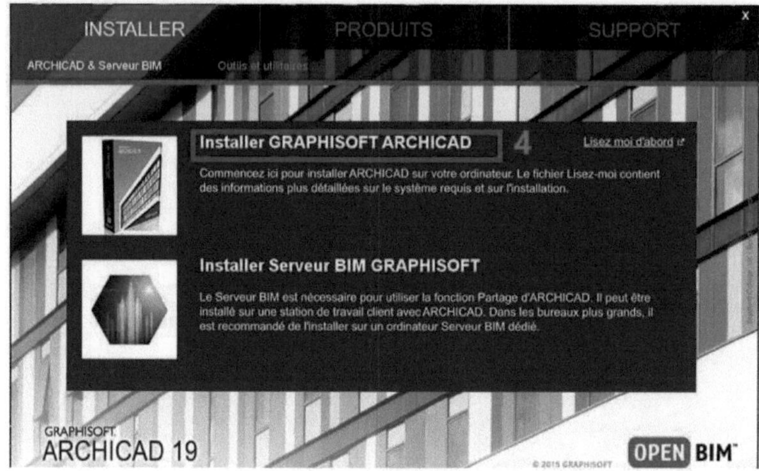

*Figure 0.3: Installing GRAPHISOFT ArchiCAD 19*

Please wait while the installation window launches the installation program, finally installing the Java RunTime environment;

*Figure 0.4: Java RunTime environment installation*

On the **Welcome to the installation wizard** screen; Click **Next** to enter the basic installation wizard.

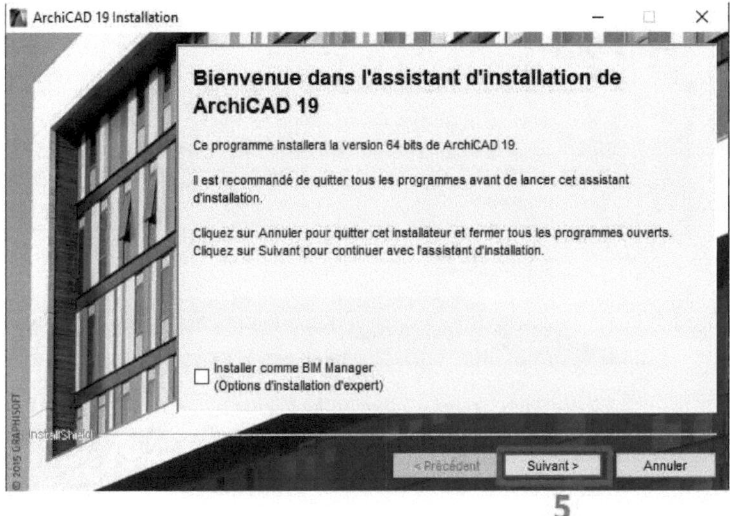


*Figure 0.5: ArchiCAD installation with basic configuration (default option)*

On the screen that appears, click **Next and** check the **I accept** box to accept the license agreement, then click **Next** again to go on to the next step, which is to choose a destination folder for the program.

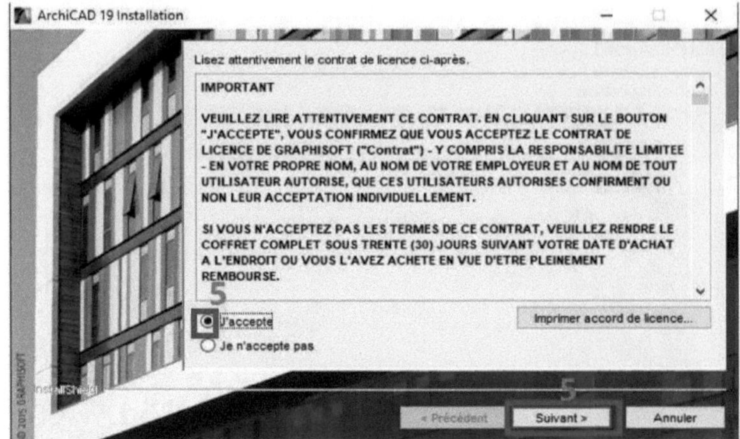

*Figure 0.6: Careful reading of the license agreement*

Click on **Next** and check the **France or Canada shortcuts** box;

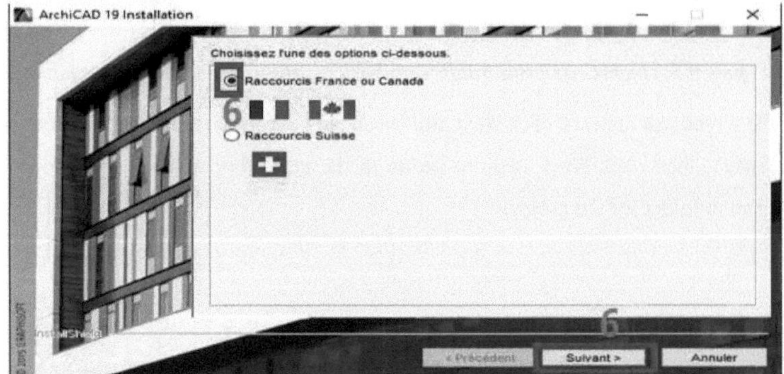

*Figure 0.7: Careful reading of the license agreement*

Click on **Next** to choose the folder in which to place your project by clicking on **Browse** and then **Next**;

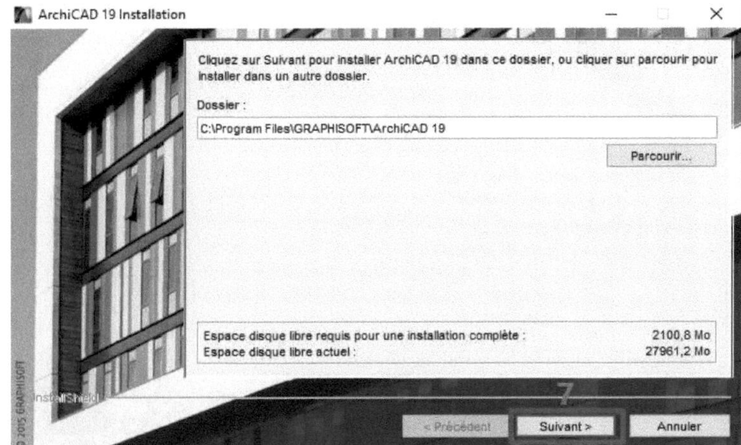

*Figure 0.8: Selection of project file*

Select **Standard** as installation type;

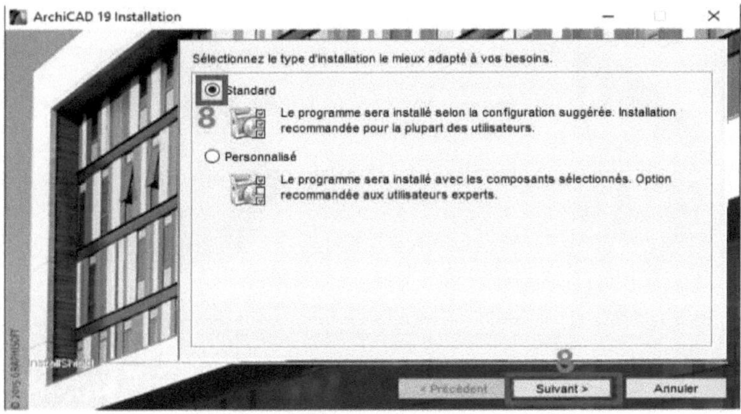

*Figure 0.9: Selecting the Standard system as the system type*

Click on **Next** and read the information about your installation;

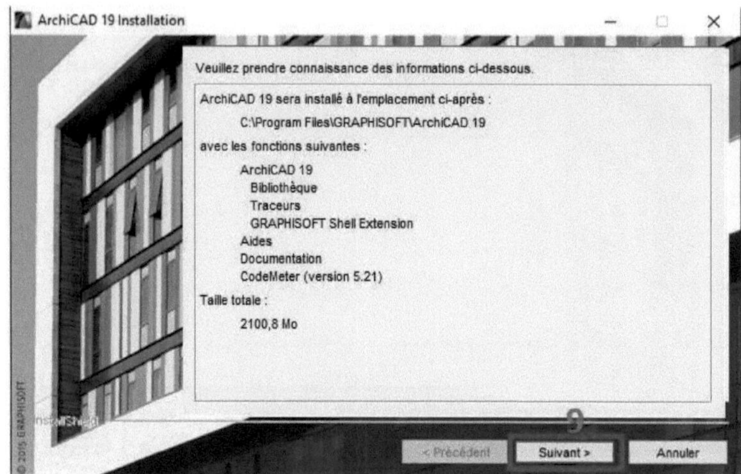

*Figure n°0.10.Reading installation information*

Click **Next** and wait until 100% of your program's components have been installed;

*Figure 0.11. The 100% clean installation*

Click **Next** to finalize the installation by **selecting one** of the **options** below;

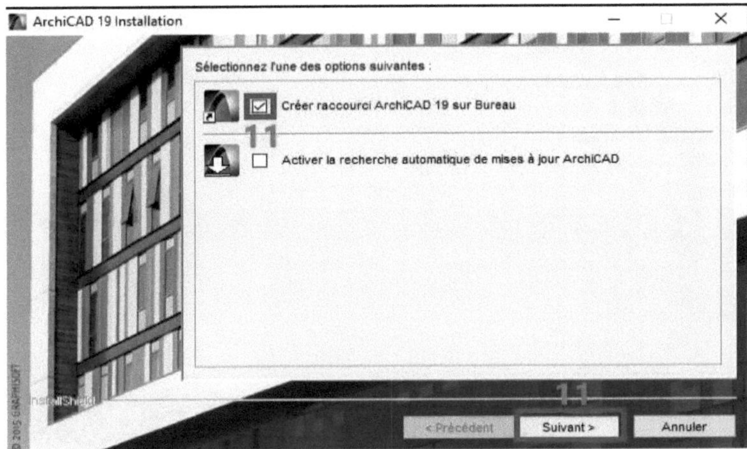


*Figure 0.12. Selecting an option to finalize installation*

To the question of whether **the installer requires you to restart your computer**, answer **yes once you**'ve saved other jobs running, and **no** to restart afterwards.

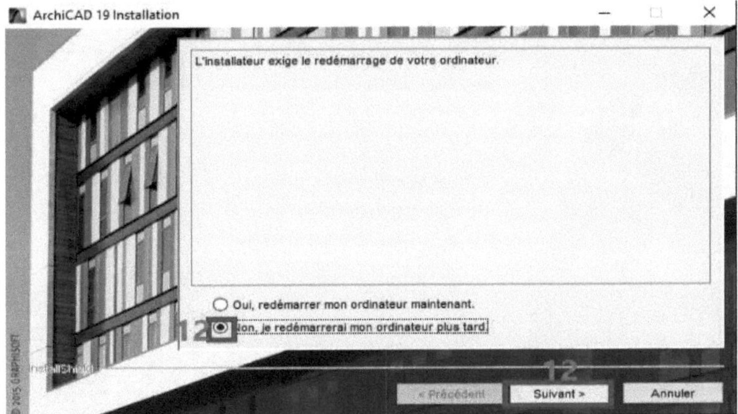


*Figure 0.13. Selecting an option to restart the computer*

Click on **Finish** to finalize installation;

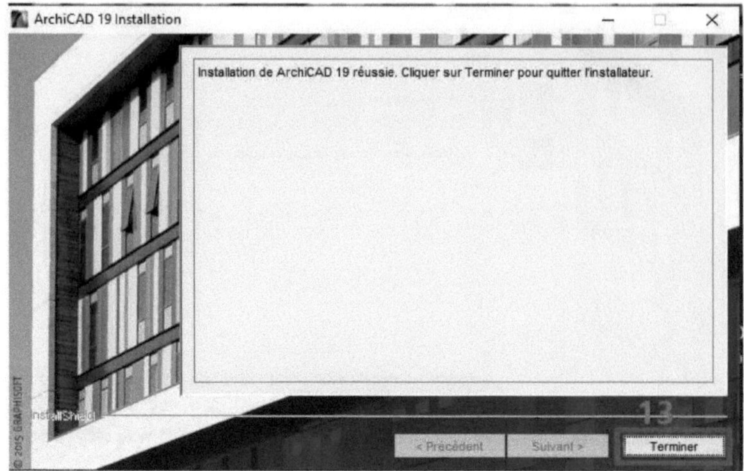

*Figure 0.14. End of installation*

After installing ArchiCAD, **launch** the program. If the program detects that you don't have a license available, a dialog box appears to help you choose the next step.

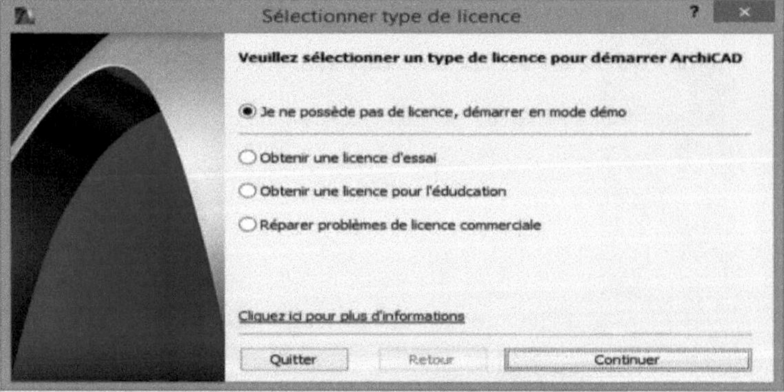

*Figure 0.15. License type selection*

You can continue in Demo mode or you can register for an educational or evaluation license by making a selection of the last 3 items in the Crack folder then do : **CTRL+C(Copy)/Disk C/Programs/GRAPHISOFT/ArchiCAD 19/CTRL V/Click on Replace file in destination** then click **on continue.**

## Help in ARCHICAD 19

To find the **help menu at the** end of the display and understand the use of a particular menu, command or software tab, use the following procedures:

**Tips :**

- ☑ **Open ArchiCAD 19** and press the **F1** function key;
- ☑ **Alternatively,** click on the **Help** menu in the running **ArchiCAD 19,** as shown in the figure below;

**Case study :**

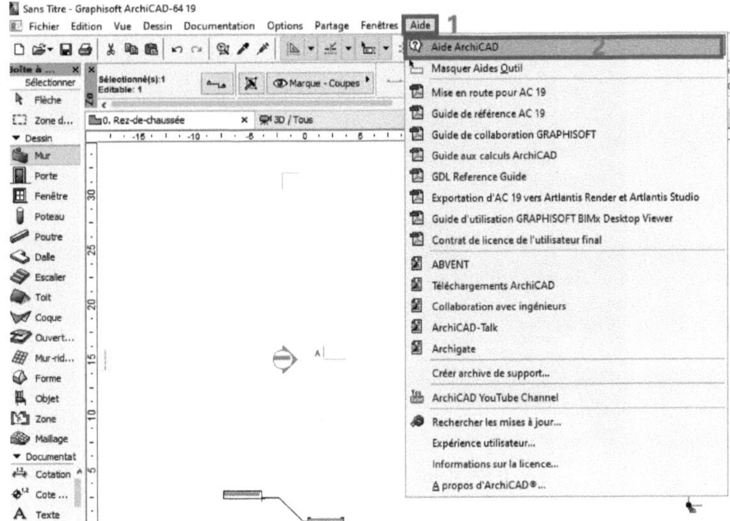


*Figure 0.15. Help display in ArchiCad 19*

# PRACTICAL GUIDE 01: OPENING AND CLOSING ARCHICAD 19

When you **start** or **open** ArchiCAD, the program automatically checks whether you have a license. If this is the case, you can start working using the following tips:

**Tips :**

- ☑ Click on **Start/All Programs/Choose ArchiCAD 19 ;**
- ☑ If the **software** has been sent to the desktop as a shortcut, **double-click** on it;

**Case study :**

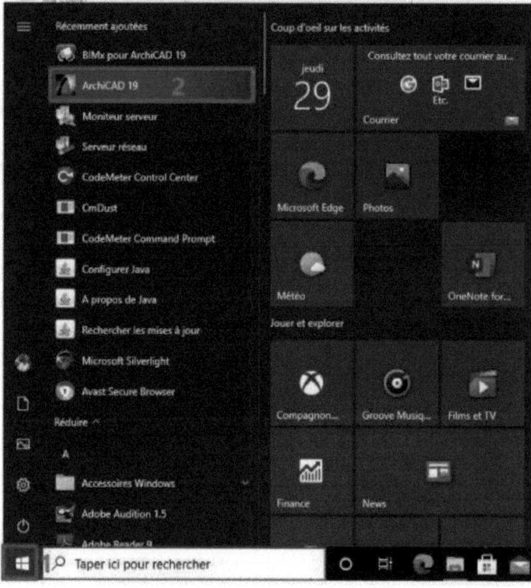

*Figure 1.0.  Open Archicad 19*

When you launch the **ArchiCAD** application, the **ArchiCAD 19 Start** dialog box appears:

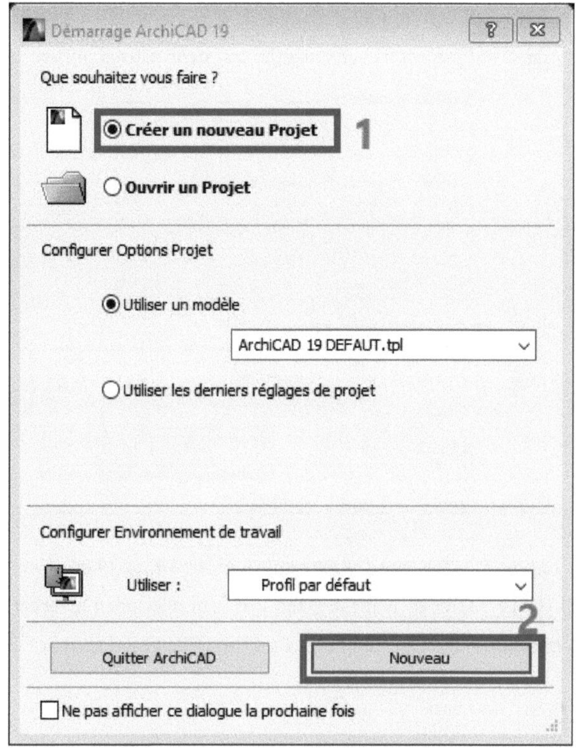



***Figure 1.1. Starting Archicad 19***

**Note:**

☑ You can also open a recent project by selecting the project name in **Workspace setup**, then **Use** and choose **Last used profile**, or from the list in the **File/Open** menu.

☑ When using ArchiCAD, your computer may experience an abnormal and unexpected interruption (crash). To avoid this, use the keyboard shortcut **CTRL+ALT+DEL/Choose Task Manager/Select ArchiCAD component/Select End of tasks**.

☑ **Good to know:** When you start Archicad, you can choose whether or not to resume your saved project by selecting the **last profile used** option.

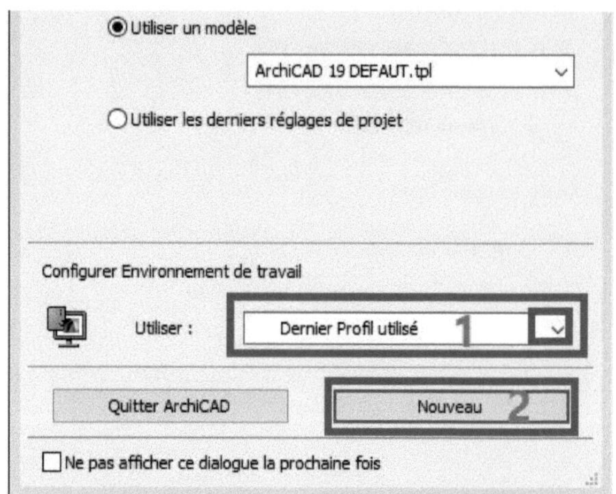

***Figure 1.2. Tips used at startup when ArchiCad 19 crashes***

Once AchiCAD 19 has been launched, the user can also open a recent project by selecting its name from the list, using tricks such as: **File/Open menu** ;

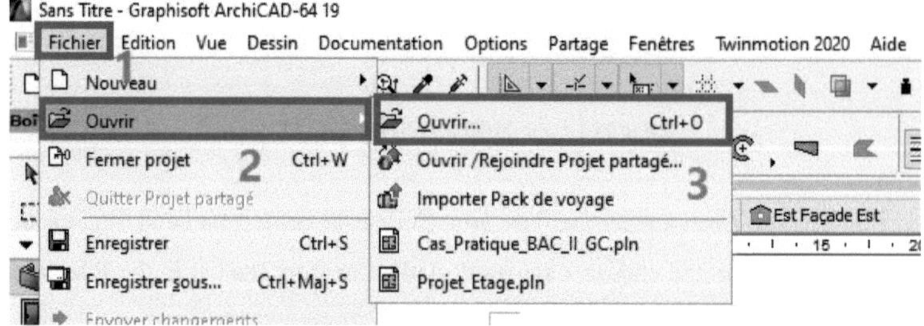

***Figure 1.3: Opening a recently used project in Archicad 19***

## ARCHICAD 19 CLOSURE

**Tips and case studies:**

- ☑ Click on the **close box** ✕ in the **current Archicad 19** window;
- ☑ Or use the keyboard shortcuts **CTRL+F4** or **CTRL+Q**;
- ☑ Or click on the **File/Exit** menu;

# PRACTICE SHEET 02: WORKING ENVIRONMENT SETTINGS

When Archicad 19 starts up, a window appears showing you various elements for starting up your drawing;

**Case study :**

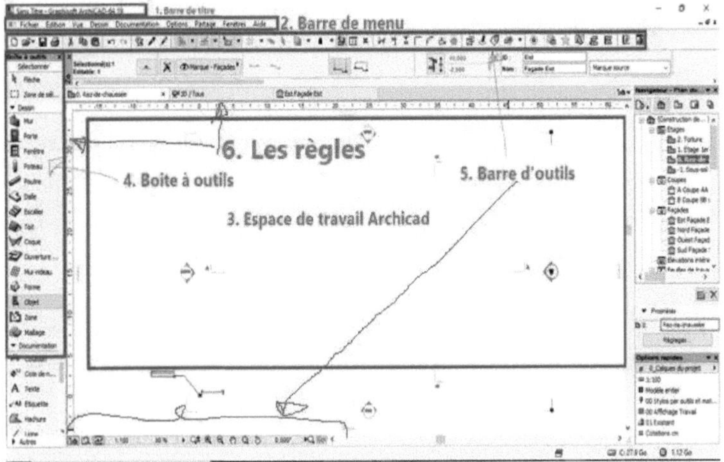


*Figure 2.0. The ARCHICAD 19 work environment*

- ☑ **Title bar**: corresponds to the actual title of your ArchiCAD project application;
- ☑ **The menu bar**: consisting of the various menus in the ArchiCAD application;
- ☑ **The workspace**: this is the main area where all drawings are made;
- ☑ **The toolbox**: Cfr Fiche pratique N°03 ;

☑ **The toolbar**: visible in the workspace, it provides tools for managing the display:

*Figure 2.1. ARCHICAD 19 toolbar display*

- ▫ to show or hide the **Navigator** palette ;
- ▫ to show or hide the **Preview** palette. **Browser** ;
- ▫ to show or hide the **Quick Options** palette ;
- ▫ to view the project's **Current Scale** ;
- ▫ Click on this icon to return to 100% display;
- ▫ to display the predefined **zoom options** ;
- ▫ to activate the wheel's **mouse zoom**;
- ▫ to **increase zoom** (x2) ;
- ▫ to **reduce zoom** (x0.5) ;
- ▫ **Crop**, to move ...

**Good to know**:

☑ It is also important to note that this toolbar can be displayed by choosing its name from the **Windows/Toolbars menu**, or by **right-clicking** on a toolbar visible on the screen to display the list of defined toolbars. Click on a toolbar name to display it.

☑ A toolbar preceded by the symbol ✔ is an activated toolbar.

**Case study :**

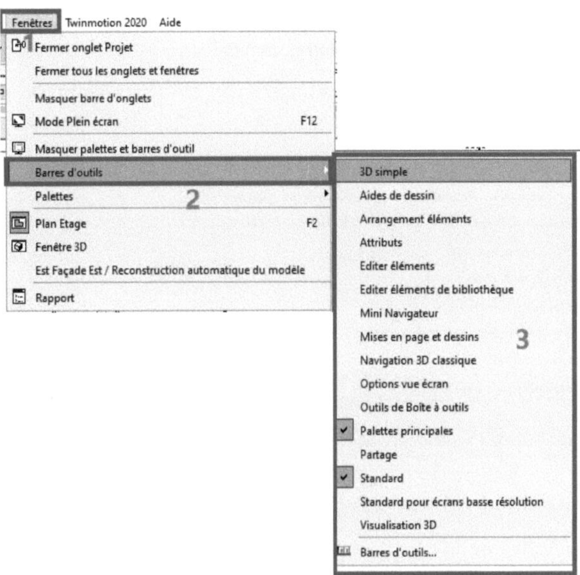



*Figure 2.2. ARCHICAD 19 toolbar display*

☑ **The ruler**: allows you to find your bearings in space according to the position of the relative origin . The relative origin can be modified by clicking on any point in the project. It can be used at any time to recreate points with coordinates 0,0 to facilitate creation. To display the ruler, click on the icon in the **Standard** toolbar or select the **View/Ruler** command. To hide it, use the same actions or right-click on the ruler and select the **Ruler** option.

In the end, for this practical sheet, the user can create several profile configurations depending on the working method used, with the following tips:

**Tips :**

☑ Click on the **Options** menu**;**

☑ Select the **Working environment** command **;**

☑ Then select **Work environment** ;

**Case study :**

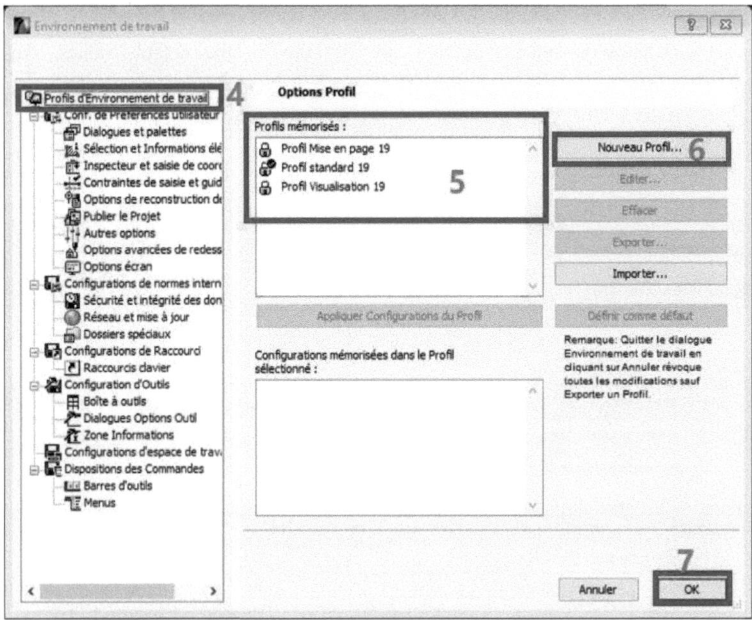

*Figure 2.3. Tips and procedures for configuring the working environment in*
*ARCHICAD 19*

For example, users can use the above dialog box to create their own profile by clicking on **New profile**, or to select one of the three default profiles.

To create a **New Profile**, use the following tips:

- ☑ Click on the **New Profile** button;
- ☑ In the dialog box that appears, configure the **New profile**;
- ☑ In the **Enter Profile Name** field, enter the name of the new profile. Each profile must have a unique name;
- ☑ In the **Select Profile Configurations** list, select each configuration type and choose the desired configuration from the list on the right. To exclude a

configuration type, select **Undefined** from the list, so that the settings of this configuration remain as they are or take on the default values. Any configuration of the new profile in **Custom** state will be stored among the configurations with the same name as the profile, and will be saved with the new profile;

☑ Click on **OK** to confirm the creation of the new profile;

**Case study :**

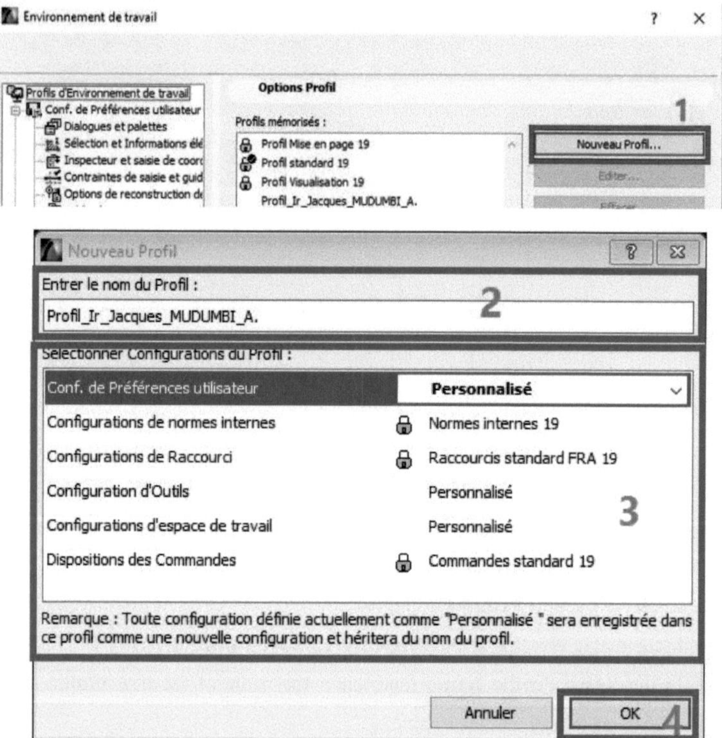


*Figure 2.4. Tips and procedures for creating a new profile in ARCHICAD 19*

**Good to know**: You can create a different profile for each user or project type.

After configuring the new profile, the user can also move on to **setting the** work **units,** using the tips below:

**Tips :**

- ☑ Click on the **options menu**;
- ☑ Select **project preferences** ;
- ☑ Select **Work units** ;
- ☑ Select **Unit Model/then** choose **Centimeters** ;
- ☑ Click on OK ;

**Case study :**

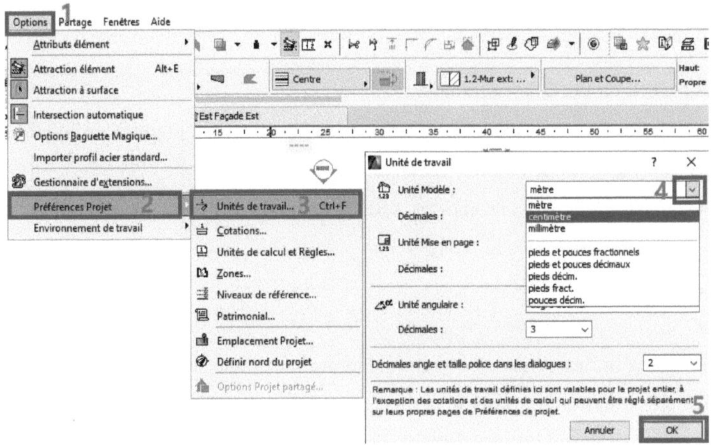


*Figure 2.5. Setting the work units*

**Note:**

- ☑ Other settings, such as **dimensioning, calculation units and rulers, and zones,** can be used using the same tricks. What's special is that for settings related to **dimensioning**, the user has to click on **"Save"** to **replace** the setting preferences once finished. For the rest, it's a matter of changing the numbers according to the settings of one user or another;

☑ It's also important to point out that before starting with the **new project** and/or drawing of your building, always start by **saving** it as a **template** to be used in ArchiCAD, for which the following tips are important for this nth remark:

- Click on the **File** menu;
- Select **Save as ;**
- **Type the name of** your project ;
- Choose **the location for** your project;
- Then click on **Save** ;

Now let's define the layout of our building with the following tips:

**Tips :**

☑ Click on **the Drawing menu;**

☑ Define the floor plan of your building by selecting an option according to your choice for most cases Select **Define floor** ;

☑ In the dialog box that appears, select **the height and altitude** of the first floor, first floor and second floor;

☑ Then click on **OK** ;

**Note:**

☑ Click on **Insert above** if you want to insert another floor;

☑ Click on **Delete storey** if you want to draw a house with only one storey;

**Case study :**

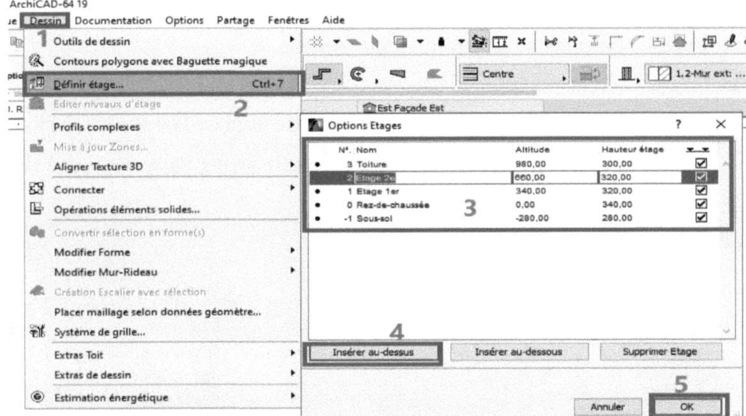


*Figure 2.6. Tips for defining the layout of your project, including the different levels and the roof.*

Before moving on, let's make a few adjustments using the drawing tool, and then start with the drawing of our building;

**Tips and tricks:**

☑ Click on a **drawing tool** ;

☑ Select and **double-click** on the **Wall, for example**;

☑ In the dialog box that appears, **change the settings as required** and start drawing;

**These last tips will be developed in detail in Practical Sheet No. 03;**

## PRACTICAL INFO 03: TOOLS AND PALLETS

Object modeling tools (mesh for terrain, wall, slab, column, beam, roof, shell, staircase, form, door, window...) are complemented by complex tools, such as curtain walls or railings, which organize several objects into a configurable system. Other 2D and 3D objects written in GDL, predefined and configurable, are available in an integrated library: structure, furniture, lamps, vehicles...

**Tips :**

When Archicad 19 starts up, all the **drawing tools are** displayed **on the left**. You can also display these tools by clicking on the **Drawing/Drawing tools menu and then choosing a tool** of your choice to start drawing.

**Case study :**

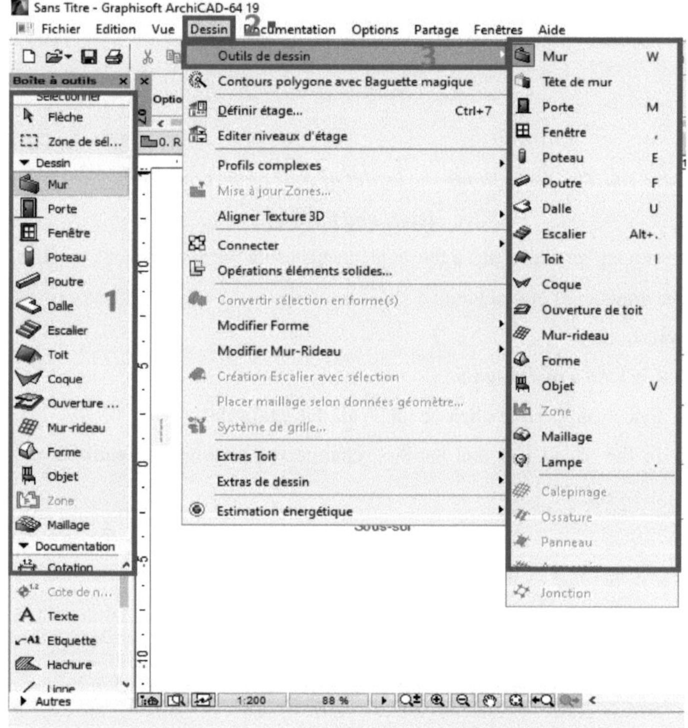


*Figure 3.0. ARCHICAD 19 tools display*

# 3.1. THE WALL TOOL

There are several types of wall in ArchiCAD: straight, curved, trapezoidal and polygonal. They can be made of a single material or several (composites), and can be complex (with Profile).

**Turnkey :**

A **wall** is a solid, vertical structure. Generally man-made, it is composed of an assembly of materials such as stones, bricks, gyproc or plaster and is intended to separate or delimit spaces;

**Tips and case studies :**

Click on the **Wall** icon in the **Toolbox** palette (**Drawing** group);

**Setting options for the Wall tool**

☑ **Once the Wall tool has been** selected, click on it and a palette appears with the **Information Zone** below:

*Figure 3.1.0: Wall drawing tool information area*

☑ To set your wall's default options, click on and a dialog box will appear, allowing you to adjust your wall's options;

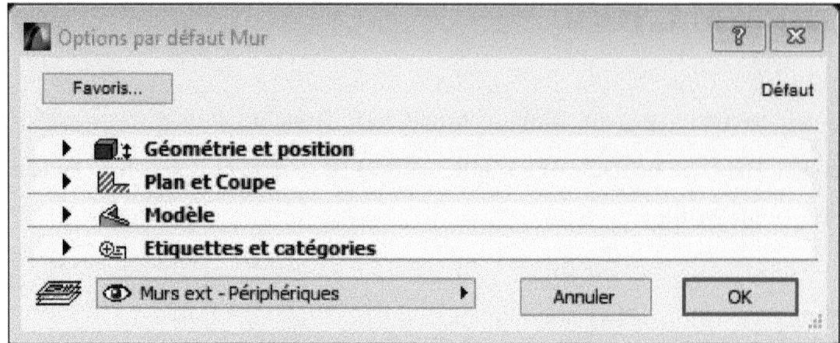

*Figure 3.1.1. Default options for the wall drawing tool*

☑ The options in the **Geometry and position** pane allow you to define the geometry and complexity of the wall;

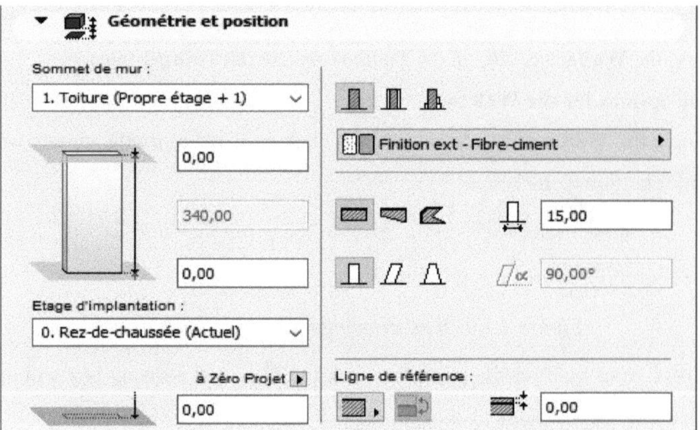

*Figure 3.1.2. Wall tool geometry and position options*

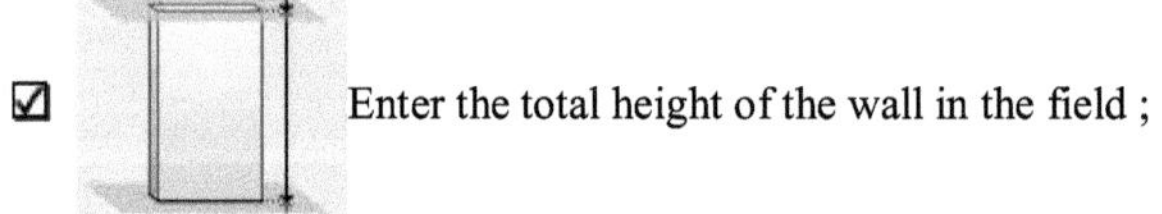 Enter the total height of the wall in the field ;

☑ In the ____ field, specify the wall altitude value. By default, this value is

calculated from the current floor (in a 3D view, it is measured from the relative

origin); to have it calculated from the **installation floor**, choose the corresponding

option in the ▸ list, then select



;

☑ In the ____ field, enter the wall elevation value in relation to the reference

level (defined via the **Options - Project Preferences - Project Levels and North**

command). This value is calculated by default in relation to **Project Zero**; to have

it measured in relation to the Level 1 or Level 2 reference, select the corresponding

option in ...

**Note**:

☑ The **plan and cut, model, label and category** options are set according to the

user's predilections...

☑ The **Model** option in most of these options is reserved for paint variations

according to your choice;

## 3.2. THE DOOR TOOL

In ArchiCAD, doors are always placed in walls and form part of the layer assigned to

the wall. As doors are real objects, a solid such as glass will be opaque in sectional

views and transparent in 3D views.

**Turnkey :**

A **door is an** opening, a structure (made of wood or metal), in a wall allowing entry

to, or exit from, an estate, a building or to move around its rooms.

**Tips and case studies :**

Door geometry can be set independently for each door type;

Click on the **Door** tool ▊ in the **Toolbox**;

**Door tool option settings**

☑ **Once the door tool has been** selected, click on this tool ▊ and a palette is displayed with the **Information Zone** below:

*Figure 3.2.0. Information area of the door drawing tool*

☑ To set the door's default options, click on ▊ and a dialog box will appear, allowing you to set your door's options;

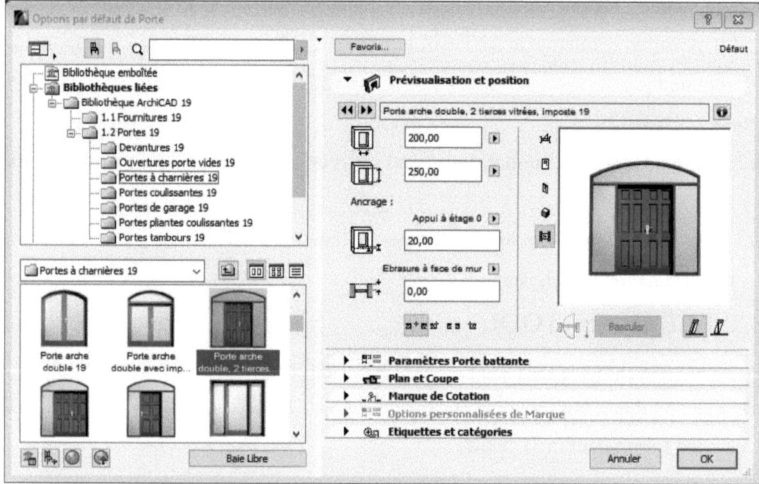

*Figure 3.2.1. Default options for the door design tool*

☑ The **Door** and **Window** tools have similar features and settings. Setting options are divided into 9 panes. To hide or show the left-hand pane, click on ▼ to the left of the **Favorites** button;

☑ To change the pane display type, click on one of the icons in the tool list;

☑ To select another door in the Library Manager or in another file, click on the **Load other object button** and select the **Open Library Manager** or **Open File dialog** option;

☑ Door object files are located by default in C:\Program Files\Graphisoft\ArchiCAD 19\Library under the name **Objects 19 Library**. Libraries from previous versions can be used. ...

☑ The **Model** option in most of these options is reserved for paint variation according to your predilection;

## 3.3. THE WINDOW TOOL

In ArchiCAD, windows behave in the same way as doors. They are always placed in walls and form part of the layer assigned to the wall.

**Turnkey :**

A **window** is an opening in a wall or sloping roof, with or without panes of glass. It differs from a door in that it does not go all the way down to the floor, but to the sill;

**Tips and case studies :**

☑ Each window style can be set in the **Information Zone** palette;

To create a window, click on the **window tool** in the **Toolbox**.

**Setting window tool options**

☑ With the window design tool displayed, click on for different settings, and an **Information Zone** palette appears;

*Figure 3.3.0. Window drawing tool information area*

☑ The proposed parameters are similar to those for the wall (see previous section);

☑ Click on the ⊞ tool to display the default options;

☑ The various window templates are displayed in the left pane of the dialog box;

*Figure 3.3.1. Default options for the window design tool*

To move a window, use the same methods as for moving a door;

**Note**:

☑ The architect can choose a window and other setting options according to his or her preference;

☑ The **Model** option in most of these options is reserved for paint variation according to your predilection;

## 3.4. THE POTEAU TOOL

Posts can be composed of two parts, the core and the cladding (plaster, decoration); their shapes can be rectangular, circular or complex (Profile). They can be incorporated into walls, either vertically or at an angle;

**Turnkey :**

A **column is** an element subject to vertical loads that are transmitted to it by load-bearing elements such as a slab, beam, etc., right up to the foundation;

**Tips and case studies :**

**Setting post tool options**

☑ Click on the **post** tool 🮲 in the **Toolbox**, and an **Information Zone** palette for your tool settings appears;

40

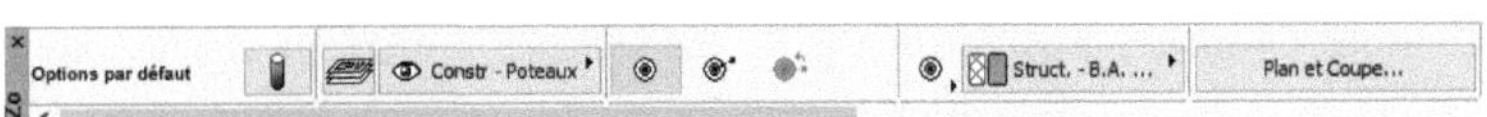

*Figure 3.4.0. Column drawing tool information area*

☑ Click on the Default **Options** tool ;

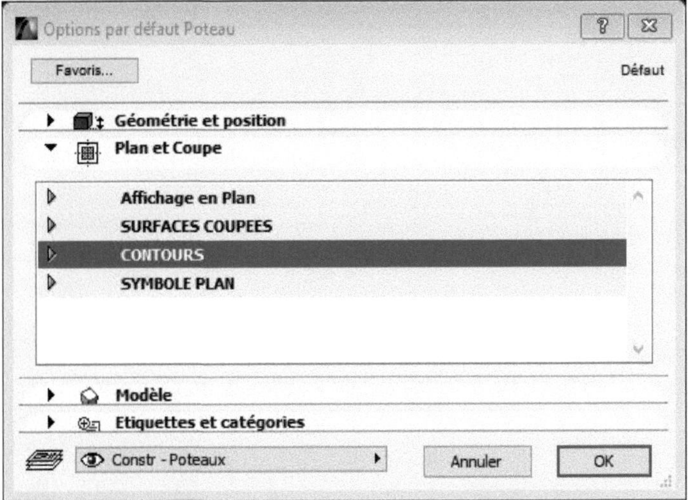

*Figure 3.4.1. Default options for the post drawing tool*

☑ Click on the **Favorites** button to use a predefined setting or to save the settings made in this dialog box;

☑ In the **Geometry and Position** pane, the architect can change certain settings according to his choice of layout;

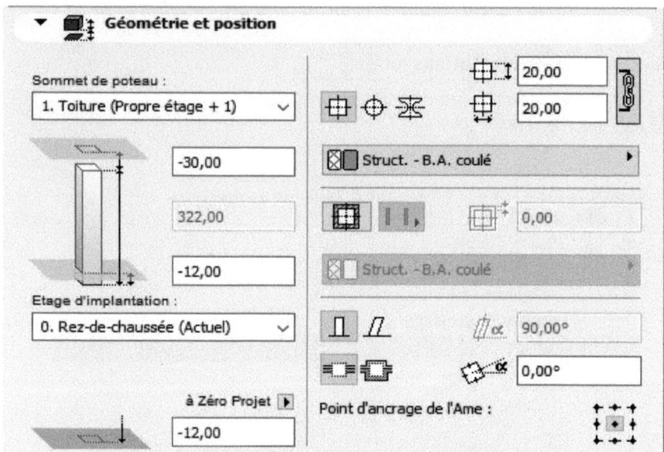

*Figure 3.4.2. Setting options for the pole drawing tool*

☑ In the     field, enter the height of the post;

☑ In the     field, enter the column height to be measured from the reference level (by default, this is the current floor). If you want the calculation to be performed from the site floor, select **A site floor** from the list in the field ;

☑ In the     field, specify the altitude of the pole base in relation to the reference level. If required, choose another reference level from the list in the field ;

☑ To rotate the post cross-section on its axis, enter the rotation angle in the     field;

☑ If the layout floor does not have to refer to ...

**Note:**

☑ For the remaining options, the user can use the measurements as they were at the start of the project;

☑ The **Model** option in most of these options is reserved for paint variation according to your predilection;

## 3.5. THE STEEL TOOL

Beams can only be horizontal constructions with clean cuts at the ends. They can be rectangular or complex (profile). You can adjust their appearance in 2D view using the **Beam** tool options;

**Turnkey :**

A **beam is** a horizontal load-bearing structure made of metal, wood or reinforced concrete, designed to support loads between and beyond its support points;

**Tips and case studies :**

**a) Setting beam tool options**

☑ To set the basic parameters, use the **Documentation/Define Model View/Model View Options** command and display the **Design Element Options** pane;

☑ **Show Beam as** options apply to all beams in the project;

☑ Choose **Whole beam** to display the entire beam, **Reference line** to display its reference lines, or **Contour line** to display only its contours;

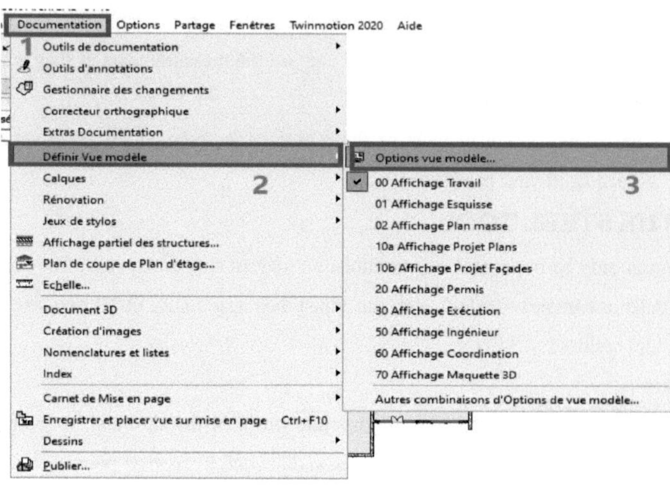

Documentation  Options  Partage  Fenêtres  Twinmotion 2020  Aide
1
Outils de documentation
Outils d'annotations
Gestionnaire des changements
Correcteur orthographique
Extras Documentation
Définir Vue modèle
Calques
Rénovation
Jeux de stylos
Affichage partiel des structures...
Plan de coupe de Plan d'étage...
Echelle...
Document 3D
Création d'images
Nomenclatures et listes
Index
Carnet de Mise en page
Enregistrer et placer vue sur mise en page   Ctrl+F10
Dessins
Publier...
2
3
Options vue modèle...
00 Affichage Travail
01 Affichage Esquisse
02 Affichage Plan masse
10a Affichage Projet Plans
10b Affichage Projet Façades
20 Affichage Permis
30 Affichage Exécution
50 Affichage Ingénieur
60 Affichage Coordination
70 Affichage Maquette 3D
Autres combinaisons d'Options de vue modèle...

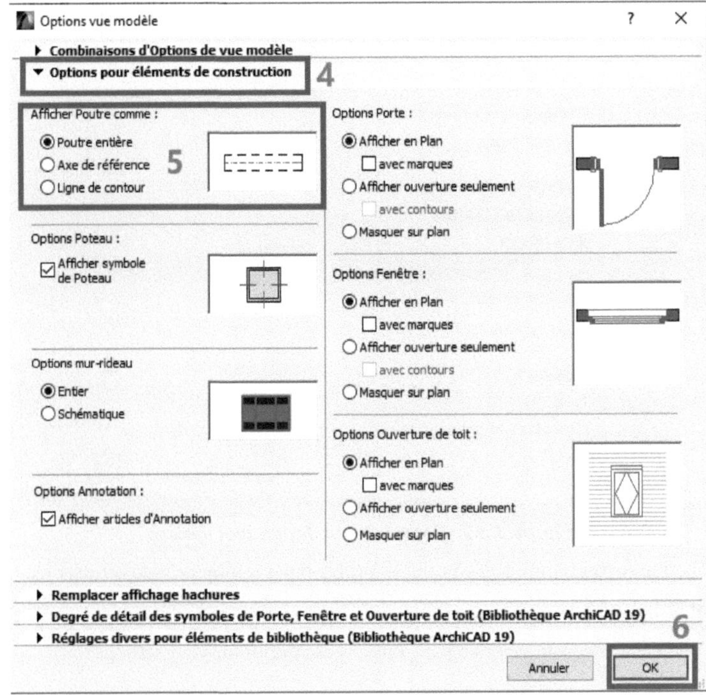

*Figure 3.5.0. Setting the basic parameters of the beam drawing tool*

## b) Modify default beam options

☑ Select the **Beam** icon in the **Toolbox**, and a dialog box appears with the **Information Area** palette;

*Figure 3.5.1: Beam design tool information area*

☑ Click on the Default **Options** tool to display the options dialog box;

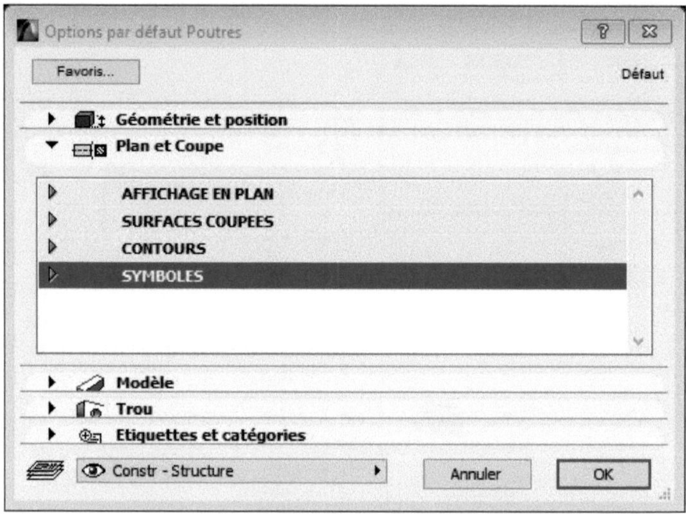

*Figure 3.5.2: Default beam design tool options*

☑ The **Favorites** button lets you select a predefined setting or save settings made in this dialog box;

☑ The Geometry and Position pane options are useful for your settings :

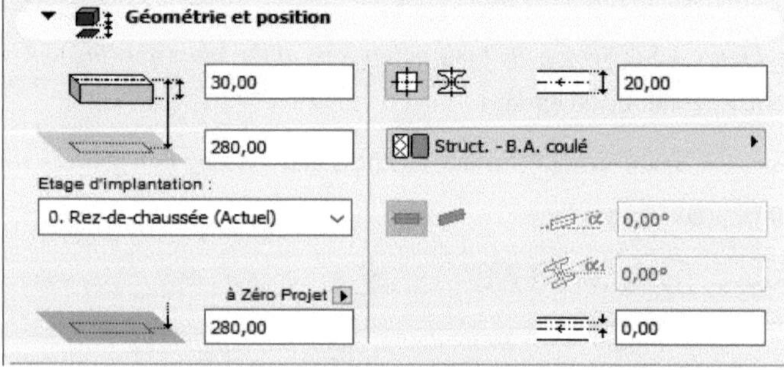

*Figure 3.5.3. Geometry and position options of the beam drawing tool*

☑ In the [  ] field, enter the height of the beam;

☑ In the [⬚] field, enter the beam altitude corresponding to the distance between the reference level ...

☑ The **Model** option in most of these options is reserved for paint variation according to your predilection;

## 3.6. THE SLAB TOOL

Slabs are used for horizontal flooring only; they cannot be used for disabled access, for example;

**Turnkey :**

☑ A **slab** is a thick layer of concrete, usually reinforced (10 cm or more), used to create a floor on a ground or upper level, over which **a screed is poured** (a few centimeters thick), on which **a covering** (tiles, carpet, vinyl, parquet or laminate) is **laid;**

☑ A **flagstone is** a set of **small slabs** or tiles laid on a screed;

**Tips and case studies :**

**Setting options for the Slab tool**

☑ Click on the **Tile** tool [⬚] in the **Toolbox** to display the **Information Zone** palette below:

*Figure 3.6.0: Slab drawing tool information area*

☑ Click on the Default **Options** tool [⬚] ;

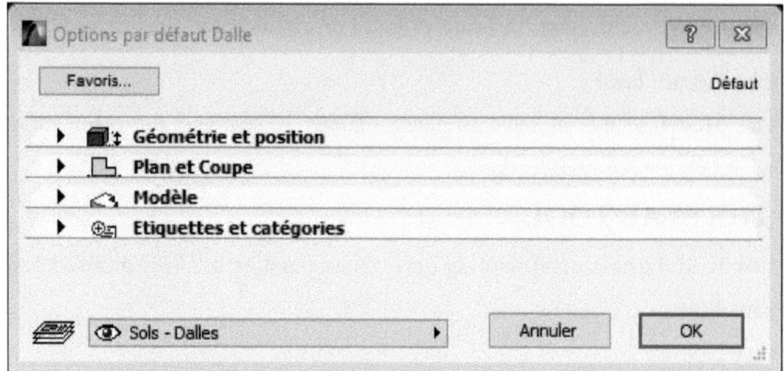

*Figure 3.6.1. Default options for the slab drawing tool*

Let's add a special feature to the **Geometry and Position pane;**

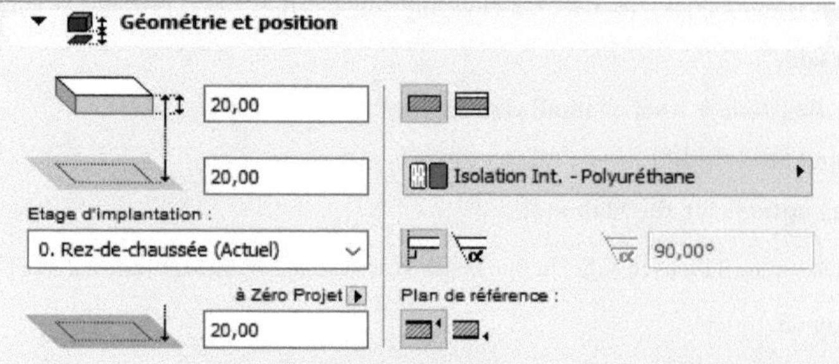

*Figure 3.6.2. Geometry and position options of the slab drawing tool*

☑ In the field, enter **the** slab **height**;

☑ In the field, enter **the** slab **height**, which corresponds to the distance between the reference level and the top of the slab (by default, this is the current floor). If you want the calculation to be based on the current floor, select **Floor** in the field list;

☑ In the field, specify the slab's **altitude in** relation to the reference level. Choose another reference level from the list in the field ;

☑ **Good to know** : If you don't want **the Layout floor to** refer to the current floor, select the **Select floor** option in the Layout **floor list** and double-click on one of the project floors;

**Note**:

☑ In the **Plan & Section** option, **PLAN VIEW** section and the **View on Floors** option list, select the **Floor** ...

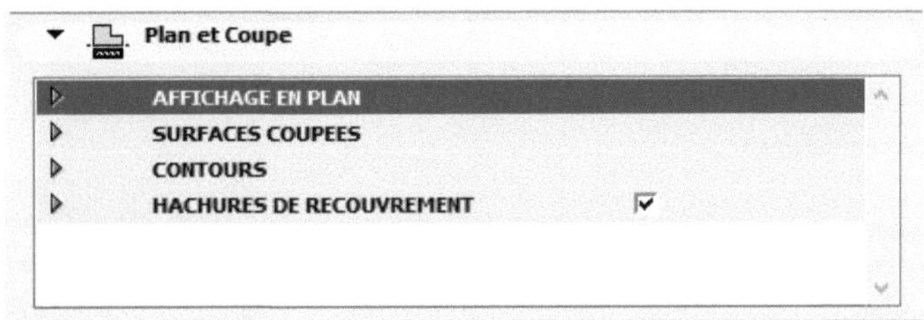


*Figure 3.6.3. Default Plan and Section options for the Slab drawing tool*

☑ The **Model** option in most of these options is reserved for paint variation according to your predilection;

## 3.7. THE STAIRCASE TOOL

The **Stairmaker** tool is used to place staircase-type library elements in the project. Although predefined staircases are available, architects can also create their own staircases using the **Stairmaker** function. Whatever the type of staircase, the options are set in the same way;

**Turnkey :**

A **staircase is** a **construction** made up of steps called degrees, which provide access to a higher floor. The word comes from the Latin **scala**, meaning **ladder**;

**Tips and case studies :**

**Setting staircase tool options**

☑ Click on the **staircase** tool ![icon] in the **Toolbox** to display the **Information Zone** palette;

*Figure 3.7.0: Staircase drawing tool information area*

☑ Click on the **Default Options** tool ⬛ ;

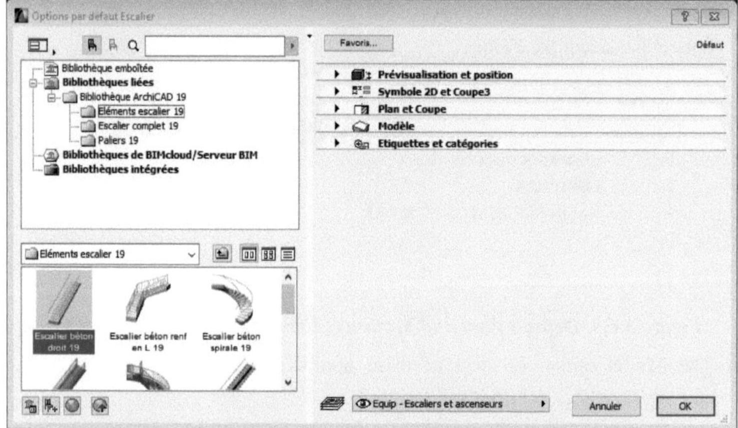

*Figure 3.7.1. Default options for the Staircase drawing tool*

The features visible in the left-hand pane are similar to those of **Windows** and **Doors**.

Let's place the milestone on the **Preview and Position pane;**

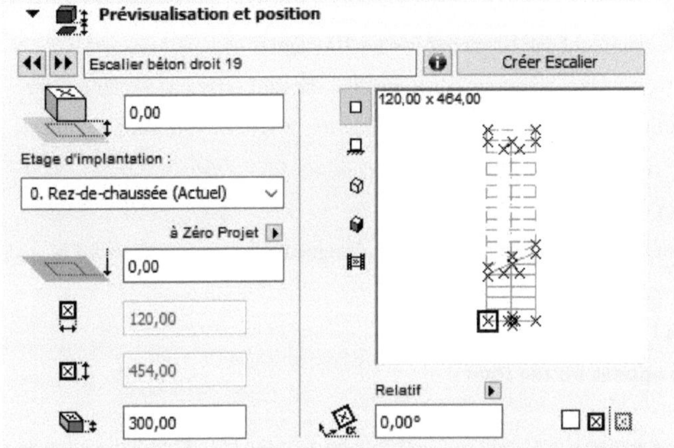

*Figure 3.7.2. Preview and position options for the Staircase drawing tool*

☑  The **Create Staircase** button opens the staircase editor;

☑  You can also select an item from the library by clicking on ◀◀ or ▶▶ to display the previous or next item respectively;

☑  In the [ ] field, specify the value of the staircase **altitude**, measured between the reference level and the top of the staircase;

☑  In the [ ] field, specify the value of the staircase **altitude**, measured from the reference level;

☑  In the [ ] field, specify the 1ère staircase dimension;

☑ The **Model** option in most of these options is reserved for paint variation according to your predilection;

## 3.8. ROOF TOOL

The   **roof** tool can be used to create 3D shapes.

The horizontal line you draw when creating the roof serves as a reference line for the roofs attitude. In most cases, this is the gutter height of your roof. It will not appear on your compressed plans;

**Turnkey :**

The **roof is** the upper part of a building designed to protect it. The **roof is** made up of a covering and a framework;

**Tips and case studies :**

**Setting options for the Roof tool**

☑ Click on the **Roof** tool  in the toolbox on the left of the application, to display the **Information Zone** palette;



*Figure 3.8.0: Roof drawing tool information area*

☑ In the **Information Zone,** you can choose the type of roofing to be used;

☑ Click on the  tool to display the Default **Options** dialog box;

52

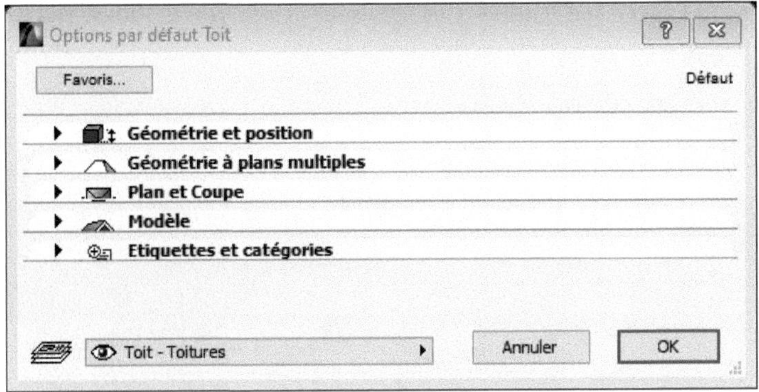

*Figure 3.8.1. Default options for the Roof drawing tool*

☑ Click on the **Favorites** button to select a saved roof model or to create a new one;

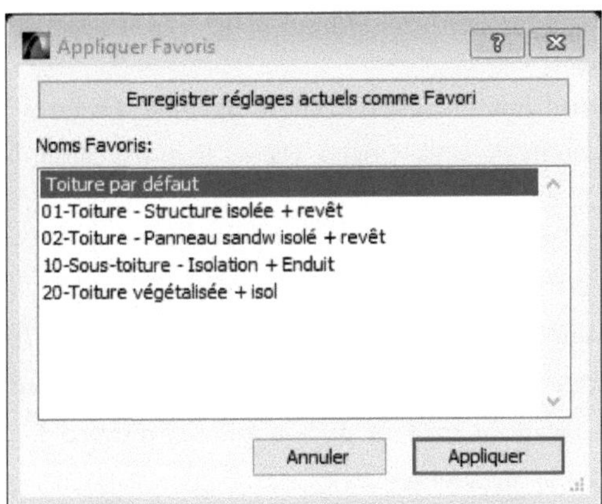

*Figure 3.8.2. Roof model selection options*

☑ The **Geometry and Position** pane allows the user to specify altitude and height values.

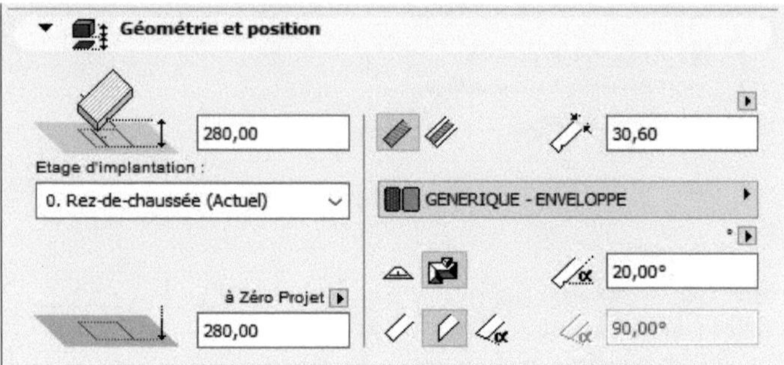

***Figure 3.8.3. Geometry and position options for the Roof drawing tool***

☑ In the ⬚ field, specify the **altitude** of the reference line. If this value is not to be calculated from the level of the current floor (in the 3D Window, it is measured from the Relative Origin), choose from the list in the **Layout floor** option ;

☑ In the ⬚ field, specify the altitude of the reference line from the reference level. By default, this value is calculated from the Project Zero. To choose another reference level, select the corresponding option from the list above ...

☑ The **Model** option in most of these options is reserved for paint variation according to your predilection;

## 3.9. HULL TOOL

In ARCHICAD, a shell is a real building element that can be used for a variety of creative purposes. Use the Shell tool to model elements, whether for the entire shell or a single object you need to customize;

**Turnkey :**

A hull is a thin, self-supporting sail with a curved surface, generally forming an arch;

**Tips and case studies :**

**Setting shell tool options**

☑ Click on the **Roof** tool in the toolbox on the left of the application, to display the **Information Zone** palette;

*Figure 3.9.0: Hull drawing tool information area*

☑ In the **Information Area,** you can choose the type of shell to be used;

☑ Click on the tool to display the **Default Hull Options** dialog box;

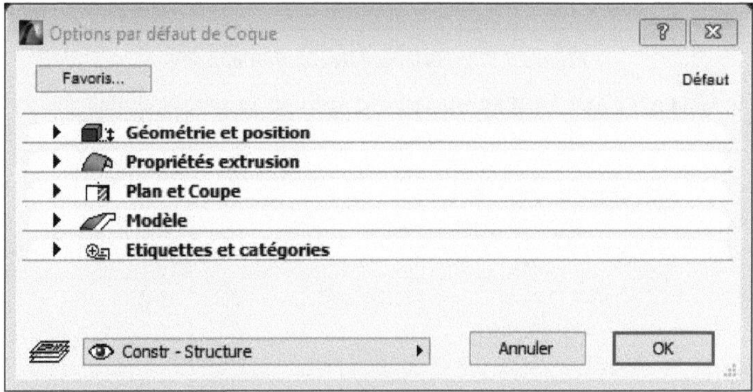

*Figure 3.9.1. Default options for the Shell drawing tool*

☑ Click on the **Favorites** button to select a saved roof model or to create a new one;

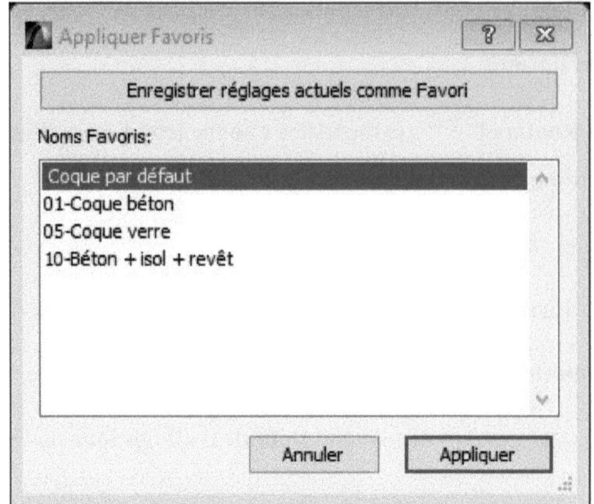

***Figure 3.9.2: Hull model selection options***

☑ The **Geometry and Position** pane allows the user to specify altitude and height values.

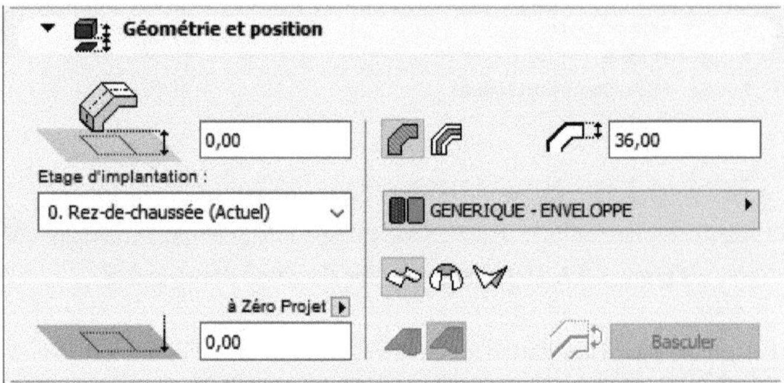

***Figure 3.9.3. Geometry and position options for the Shell drawing tool***

☑ In the [field icon] field, specify the **altitude** of the reference line. If this value is not to be calculated from the level of the current floor (in the 3D Window, it is measured from the Relative Origin), choose from the list in the **Layout floor** option ;

☑ In the field, specify the altitude of the reference line from the reference level. By default, this value is calculated from the Project Zero. To choose another reference level, select the corresponding option from the list above ...

☑ The **Model** option in most of these options is reserved for paint variation according to your predilection;

# 3.10. ROOF OPENING TOOL

**Roof openings** behave in the same way as doors and windows: when you install a roof opening, the roof is modified (creation of the opening) in the same way as when you install a door in a wall; the opening is interactive with the dimensions you specify in the parameters ;

**Turnkey** :

A **roof opening is** a window drilled into a **roof** (directly integrated or on an upstand) whose frame, made of wood or metal in one piece, opens by rotation (oscillating frame), rotation and/or panoramic projection (snuffer frame);

**Tips and case studies :**

**Setting options for the Roof Opening tool**

☑ Click on the **Open Roof** tool in the **Toolbox** to display the **Information Zone** palette;

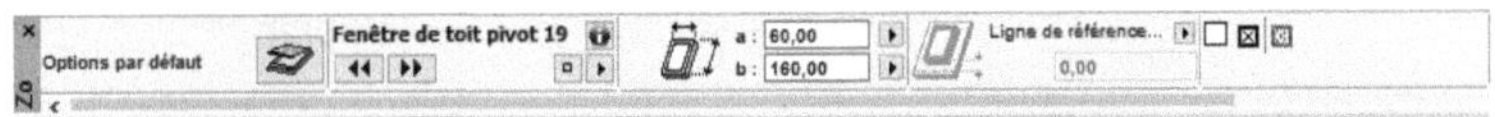


*Figure 3.10.0. Roof opening drawing tool information area*

☑ Click on the Default **Options** tool to display the **Default Roof Opening Options** dialog box.

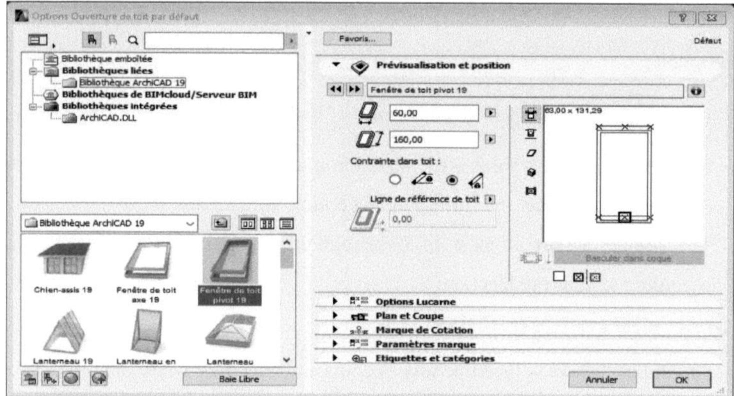

***Figure 3.10.1. Default options for the Roof opening drawing tool***

☑ **The Preview and Position pane** options allow you to set the dimensions and positions of the opening;

☑ The **Brand Settings pane** allows you to set the parameters of the roof window;

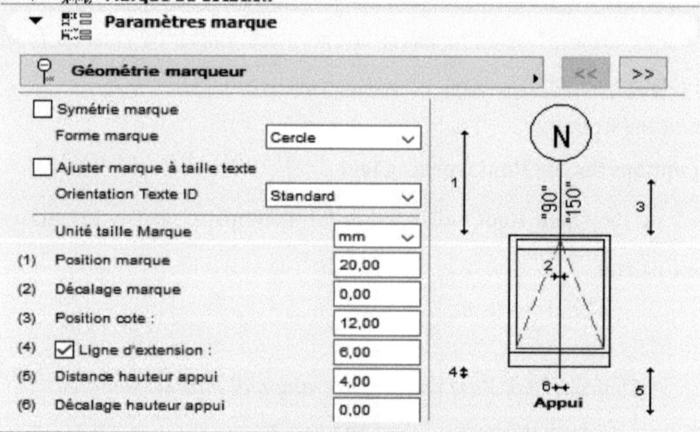

***Figure 3.10.2. Roof opening parameter setting options***

☑ **Volet Options Lucarne** allows precise adjustment of the roof window;

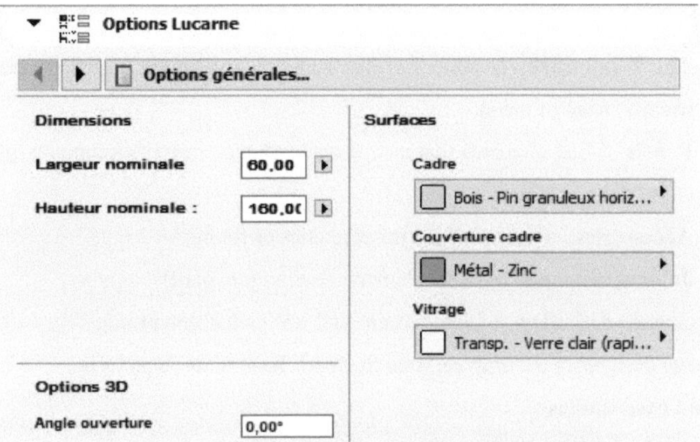



*Figure 3.10.3: Parameter setting options for skylight*

☑ The **Plan & Cut, Template and Lists & Labels panes** are the same as those for the Door, Window, Corner Window, Stair and Object-Lamp tools.

**Good to know**: Roof openings can be placed in roofs in Plan or 3D view. In Plan view, you must place the opening within the roof contour; in 3D view, when you click on the roof plane, ArchiCAD automatically places the opening in the roof plane. Roof openings behave in the same way ...

## 3.11. CURTAIN WALL TOOL

The **Curtain Wall** tool in the **Toolbox is** used to create a curtain wall in the Plan, Section, Facade, Interior Elevation and 3D windows;

**Turnkey :**

**Curtain walls** are also known as "**curtain facades**". It is a facade wall that closes off the building envelope without contributing to its stability. The panels are supported, storey by storey, on a fixed skeleton;

It can also refer to the external, non-load-bearing envelope of a building with a steel or reinforced concrete **structure** (suspended from this **structure**, the **curtain wall** is usually largely glazed and made of standard prefabricated elements, panels possibly joined by a grid);

The curtain wall is made up of :

- **The framework** (or supports) that supports the entire "light" façade; it is usually made of metal;
- **Panels**: filling elements that can be made of sheet metal, aluminium, glass or stone slabs;
- **Accessories**: elements added for aesthetics or fixing;
- **Joining elements**: optional elements used to join panels together;

The geometric definitions of the curtain wall are called **calepinage**. By extension, calepinage designates the drawing used to create the structure's patterns;

**Tips and case studies :**

**Setting options for the Curtain Wall tool**

☑ Click on the **curtain wall** tool [icon] in the **Toolbox** (**Drawing** group) and, if required, display the **Information area**;

*Figure 3.11.0: Information area of the Curtain Wall design tool*

☑ Click on Default **options** [icon]

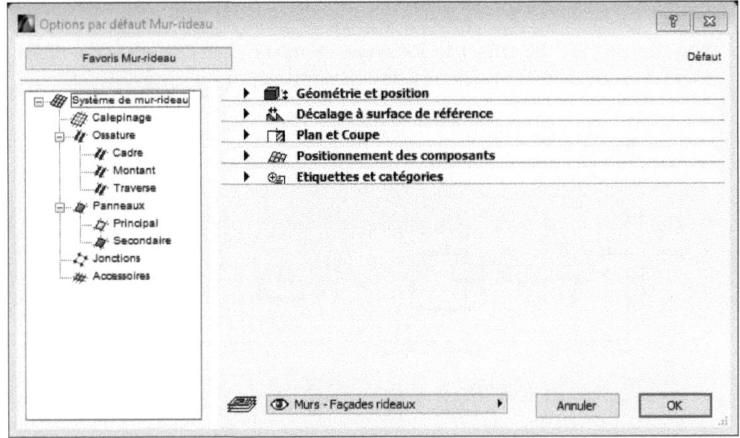

*Figure 3.11.1. Default options for the Curtain Wall drawing tool*

☑ On the left, select **Curtain wall system,** if you wish to set the general curtain wall parameters, or select the component you wish to work on;

☑ The **Geometry and Position pane is** used for various parameters;

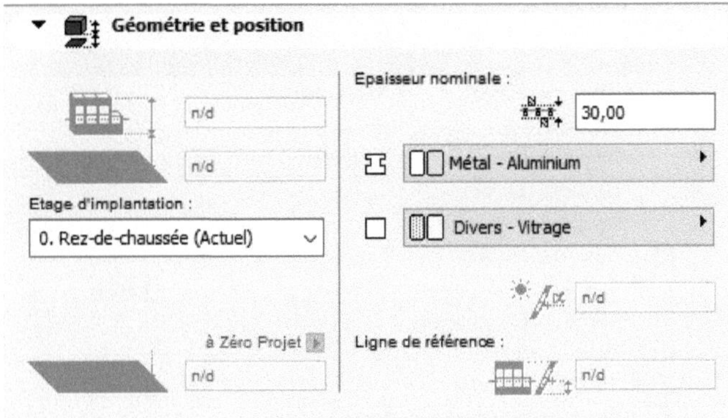

*Figure 3.11.2: Geometry and position options for the curtain wall design tool*

☑ Enter the curtain wall thickness in the **Nominal thickness** field;

☑ Select the curtain wall **floor** in the corresponding list;

The following settings are only active when modifying a curtain wall.

☑ You can also use the **Offset to Reference Surface** pane to specify the offset of frame, stiles, rails and panels from the reference surface for each value.

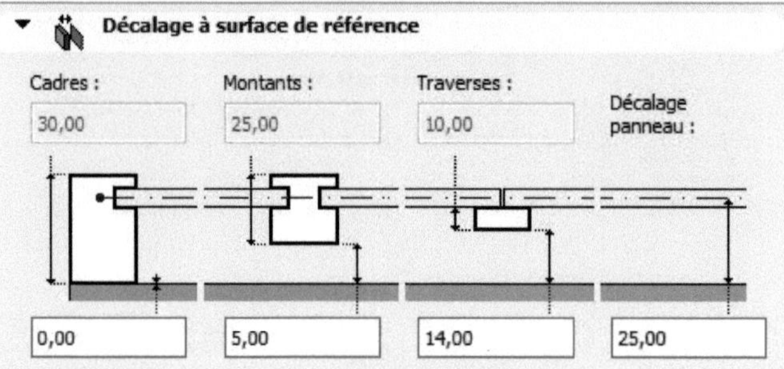


*Figure 3.11.3: Curtain wall design tool options: offset to reference surface*

☑ Use the **Plan and Trim** pane to modify, if required, the pen and hatch attributes for the elements that make up the curtain wall in the plan view, on cut elements and contours;

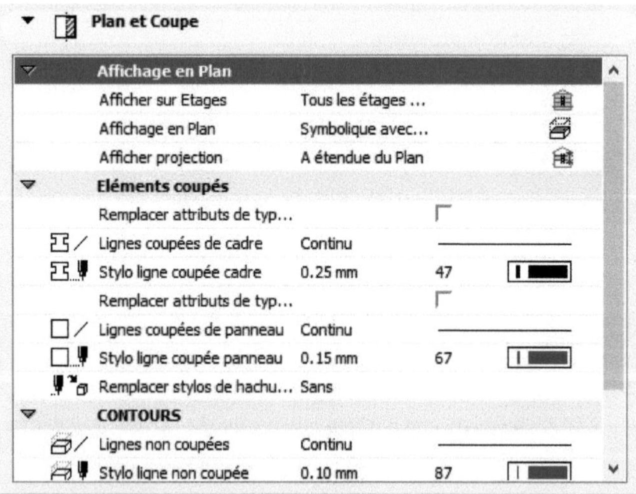

*Figure 3.11.4. Plan and cut options in the Curtain Wall drawing tool*

☑ Use the **Component Positioning** pane to choose the placement method for junctions, frames and uprights.

☑ For junctions, activate the **At all points in the layout** option;

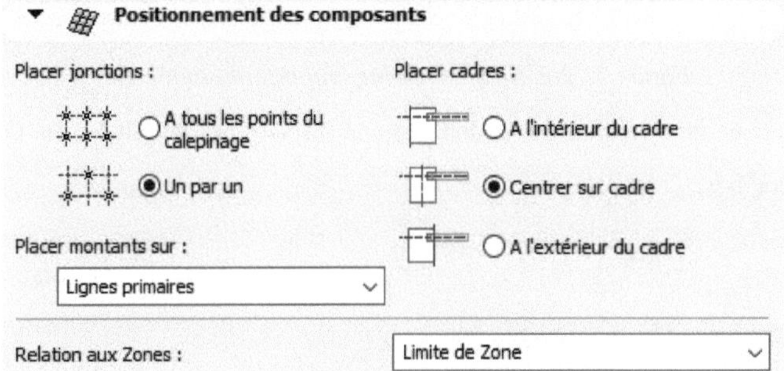


*Figure 3.11.5: Component positioning options for the Curtain Wall design tool*

Other options for editing a curtain wall are developed in **Case Study N° 04** on pages 81 to 82.

## 3.12. THE FORM TOOL

Use **the Shape tool** to create free-form elements directly in ARCHICAD. This eliminates the need to import special shapes from other applications.

**The Forme tool** is fully integrated into ARCHICAD, and operates according to a logic and interface that are familiar to you. Like other construction elements, the essential structure of the Shape is derived from its construction material.

You can activate the priority-based connection between Shapes and other element types;

**Turnkey :**

A **shape** models complex volumes;

**Tips and case studies :**

**Setting shape tool options**

☑ Click on the **Shape** tool  in the **Toolbox** to display the **Information Area** palette;



*Figure 3.12.0. Shape drawing tool information area*

☑ Click on the Default **Options** tool  to display the **Default Shape Options** dialog box.

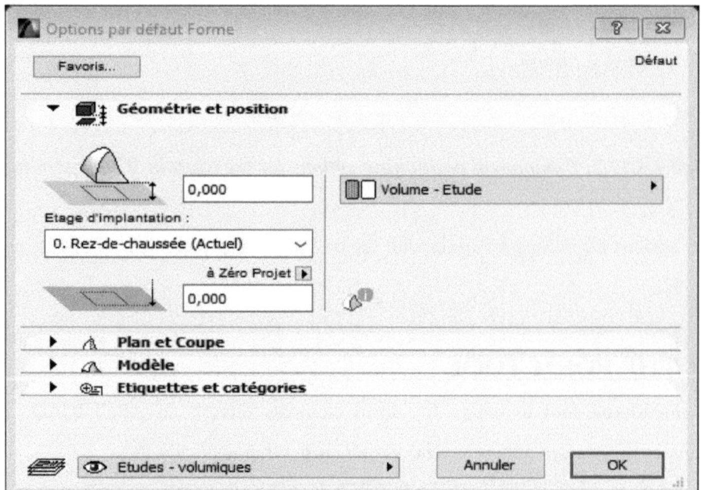


*Figure 3.12.1. Default options for the Shape drawing tool*

☑ **The Geometry and Position pane** options are used to set the dimensions and positions of the Shape ;

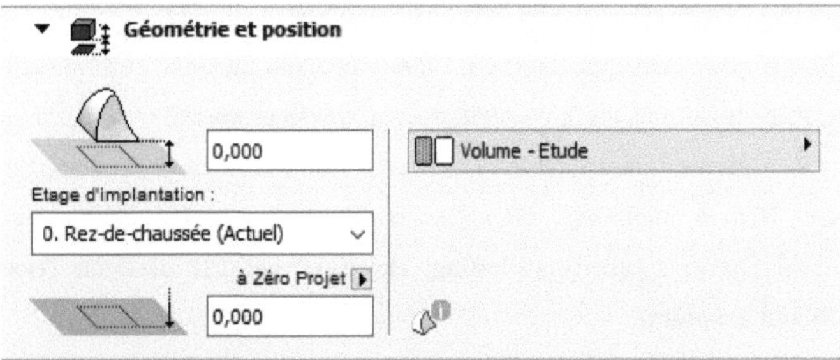



*Figure 3.12.2: Shape geometry and position options*

☑ The **Plan & Cut, Template and Lists & Labels panes** are the same as those for the Door, Window, Corner Window, Stair and Object-Lamp tools.

## 3.13. OBJECT TOOL

The **Object** tool is used to place object-type library elements in the project. Note, however, that the navigation area on the left of the **Object Options** dialog is identical to that of other library elements.

**Turnkey** :

An **Object** is any concrete thing, perceptible by sight or touch: Perception of **objects**. A solid thing considered as a whole, manufactured by man and destined for a certain use: A lamp, an armchair are **objects**. ...

**Tips and case studies :**

**Folder and subfolder navigation (Library view)**

In the Default **Object Options** dialog box, you can access a wide variety of objects that do not belong to a dedicated sub-type. These objects are divided into three main folders:

- **Basic library ;**
- **Display objects; and**

- **Extension library, as well as a large number of subfolders.**

As the folder names indicate, the Basic Library contains all kinds of furniture (**beds; chairs; office furniture, etc.**); decorations (e.g. **clocks or vases**); sanitary and leisure items (e.g. **a piano, billiard table or TV set**). Additional folders contain Building Structures (**fences, moldings, etc.**); Special Constructions (**fireplace, shutters**); Mechanical Elements (**air conditioning, elevator**); and 2D Elements (**electrical symbols and graphics**).

The Visualization folder contains site elements (e.g. **trees**) and objects representing **people and vehicles**.

**Setting object tool options**

☑ Click on **the** tool in the **Toolbox** to display the **Information Zone** palette;

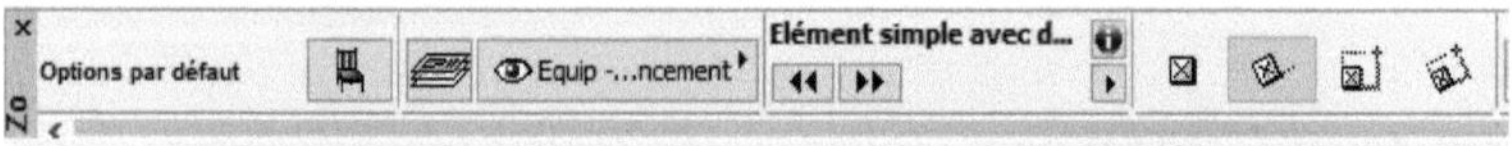

*Figure 3.13.0. Information area of the Object drawing tool*

☑ Click on the Default **Options** tool to display the **Default Object Options** dialog box.

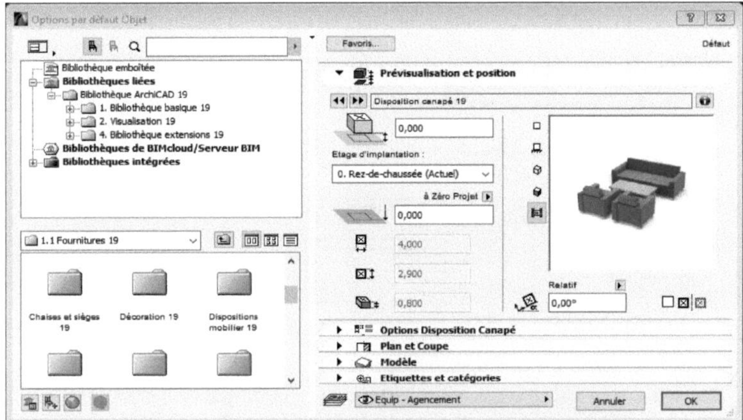

*Figure 3.13.1. Default options for the Object drawing tool*

☑ **The Preview and Position pane** options allow you to set object dimensions and positions;

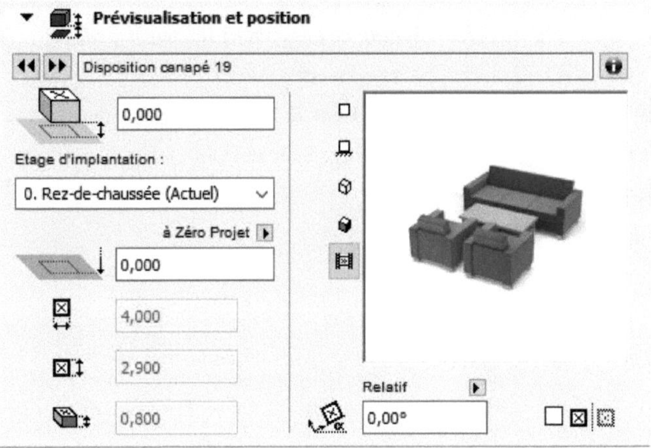

*Figure 3.13.2: Preview pane options and Object drawing tool position*

☑ The **Layout and Sofa Options pane lets you** adjust the parameters of the selected object;

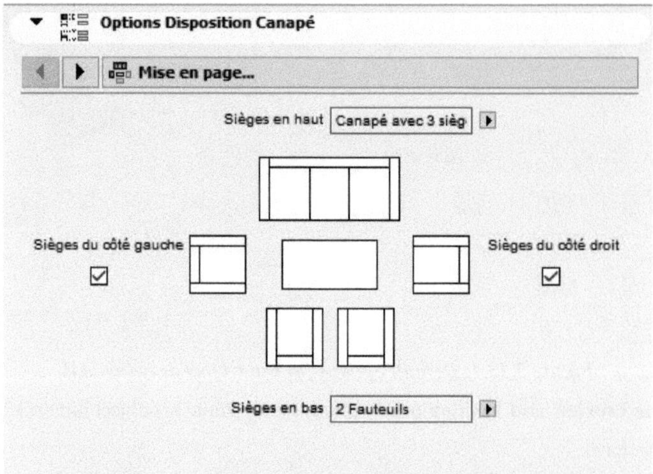


*Figure 3.13.3: Setting layout options for selected object*

☑ The **Plan & Cut, Template and Lists & Labels panes** are the same as those for the Door, Window, Corner Window, Stair and Object-Lamp tools.

# 3.14. ZONE TOOL

**Zones** are used to calculate the surface areas and volumes of your project. You can assign colors to visualize the types of space in your project. Use predefined zones, or create your own by selecting the command in the **Options/Element Attributes/Area Categories** menu (or use the keyboard shortcut **[Ctrl][Alt] Z**).

**Turnkey** :

**Zones are** 3D elements with a specific representation in plan view; they are, in fact, represented by **a hatch and a zone mark**.

**Tips and case studies :**

**Setting Zone tool options**

☑ Click on the **Zone** tool in the **Toolbox** to display the **Zone Information** palette.

*Figure 3.14.0. Drawing tool information zone Zone*

☑ Click on the Default **Options** tool to display the **Default Zone Options** dialog box;

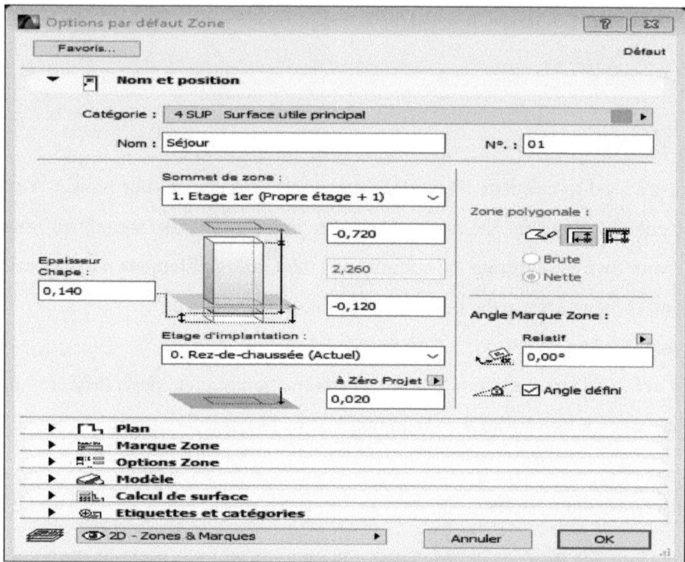

*Figure 3.14.1: Default options for the Zone drawing tool*

☑ The **Name and Position** options allow you to set the dimensions and positions of the Zone;

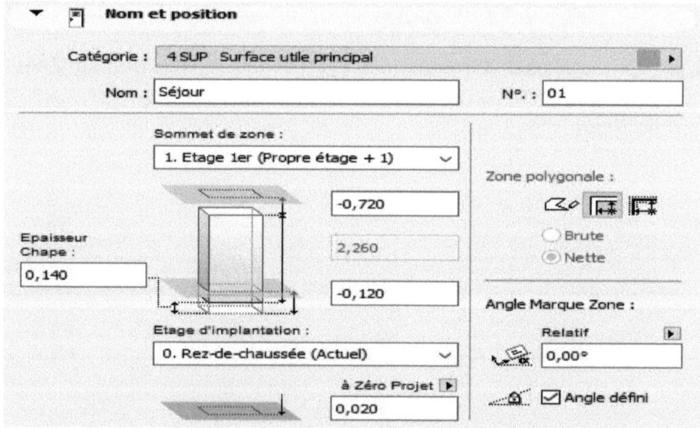

☑ In the **Zone category** list, select a zone category from the list of categories defined by the **Options/Element attributes/Zone categories** command. Each category consists of a name, a code, a color, a zone mark and a set of parameters.

☑ Enter the zone **name** and, in the **No.** field, its number: the name and number are used to identify the zone marker on the plan, and must be unique.

☑ In the **Zone height** field, enter the height of the zone body, measured from the zone's lowest point.

  - In the **Zone Level** field, enter the reference level of the zone measured from the floor elevation.

  - In the **Screed thickness** field, enter the height of the lower part ...

☑ The **Zone Mark pane** allows you to set the parameters of the selected zone;

☑ The **Zone Options pane** allows you to set options for the selected zone;

☑ The **Surface Calculation pane** allows you to set the wall, curtain wall, post and roof of the selected zone;

☑ The **Plan, Model and Lists & Labels panes** are the same as those for the Door, Window, Corner window, Staircase and Lamp-object tools.

## 3.15. THE MESH TOOL

Meshes can be used to create terrain in your project. Surfaces created in this way can be arbitrary, inclined or horizontal.

**Turnkey** :

**Meshes are** surfaces of any shape created by **defining** the altitude of their characteristic points and interpolating between these points. ... .

**Tips and case studies :**

**Setting mesh tool options**

☑ Click on the **Mesh** tool in the **Toolbox** to display the **Information Zone** palette.

*Figure 3.15.0: Mesh drawing tool information area*

☑ Click on the Default **Options** tool to display the **Default Mesh Options** dialog box;

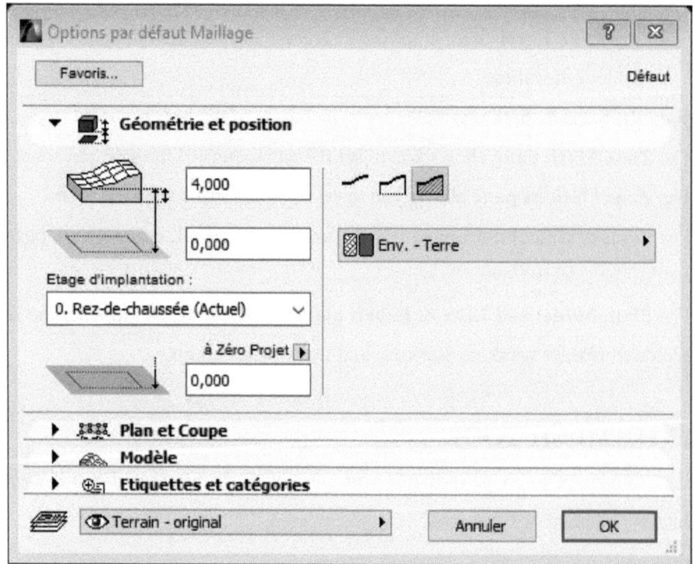



*Figure 3.15.1: Default options for the Mesh drawing tool*

☑ **The Geometry and Position pane** options are used to set mesh dimensions and positions;

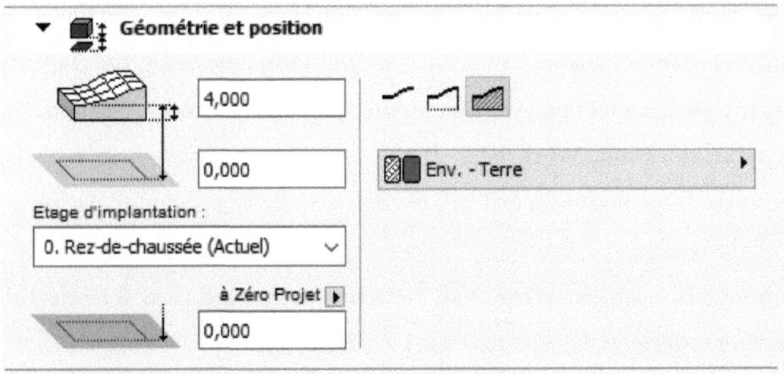



☑ In the [ ] field, specify the depth to which the mesh extends below the reference plane;

☑ In the [ ] field, define the height of the mesh measured from the reference plane. By default, this reference level is the current floor level;

☑ In the field [ ], define the height of the mesh measured from the reference level;

☑ Click on the icon corresponding to the desired geometric mesh construction method:

- to create simple surfaces with the **Mesh** tool;

- to create skirted surfaces with the **Mesh** tool;

- to create solid bodies with the **Mesh** tool;

☑ To select a floor other than the current one, in the **Floor** list, select the **Select floor** option and double-click on one of the project's floors.

☑ The **Plan & Cut, Model and Labels and Categories panes** are the same as those for the Door, Window, Corner Window, Stair and Lamp Object tools.

**Good to know:**

☑ The tools grouped together on the left-hand side of **Archicad** in the **Documentation** menu in the toolbox, such as (**Dimensioning, Level dimension, Text, Label, Hatching, Arc/Circle line, Polyline, Drawing, Section, Facade, Interior elevation, Worksheet, Detail and** Change) are support tools and are only used when necessary and will be dealt with in the auditorium during tutorials with the teacher;

☑ Other tools, such as **Arrow and Selection Area**, are used for **selecting** and **moving objects** in the Archicad work environment;

☑ Other tools such as **Grid Element, Wall Head, Corner Window, Lamp, Pencil Dimension, Angular Dimension, Spline, Hot Spot, Figure and Camera** are also tools for supplementing the drawing and are only used in the same way as other tools such as **Door, Window, Corner Window, Staircase and Object**;

Finally, let's say that **ARCHICAD's palettes** help you **build, modify and find** your elements. You can show or hide each palette separately using the **Windows/Pallets** command.

**Turnkey :**

A **palette**: is either an element of a **computer** program's graphical interface, used to select from a limited number of choices, not necessarily colors (for example: a tool **palette**). ...

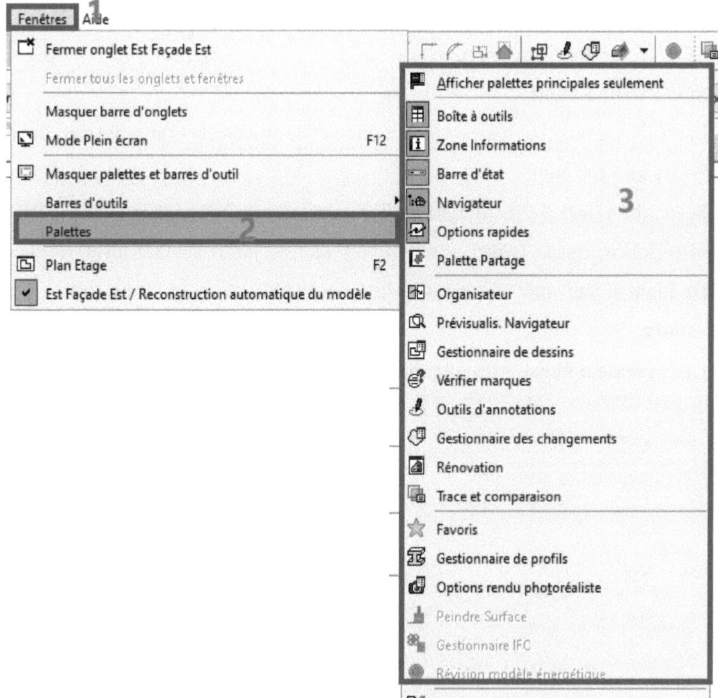



*Figure 3.1. ARCHICAD 19 Palettes display*

**Note:**

The main palettes (**Toolbox, Information Area, Status Bar, Navigator and** Quick **Options**) can be activated simultaneously by selecting the **Windows/Palettes/Display main palettes only** command.

# PRACTICAL INFO 04: WALL-POST-SLAB-ROOFING-ROOFING-PHOTO RENDERING

As hinted at in **Practical hint no.02,** once you've set your working units and defined the level of your building, start by using the **Wall** drawing tool, developed further in **Practical hint no.03.** To do this, the following tips can be used first:

**Tips :**

☑ Click on the **Drawing** menu;

☑ Select **Define Floor**;

☑ In the dialog box that appears, select and define **the levels that will contain your house and the roof**;

☑ Select the values for the **height and altitude** of your house, then click on **OK** ;

**Good to know**: Select **Insert above** if you want to insert another level, or choose **Delete Floor** if you want a house with just one level;

**Case study:**

☑ **Let's create a house with only one level (first floor);**

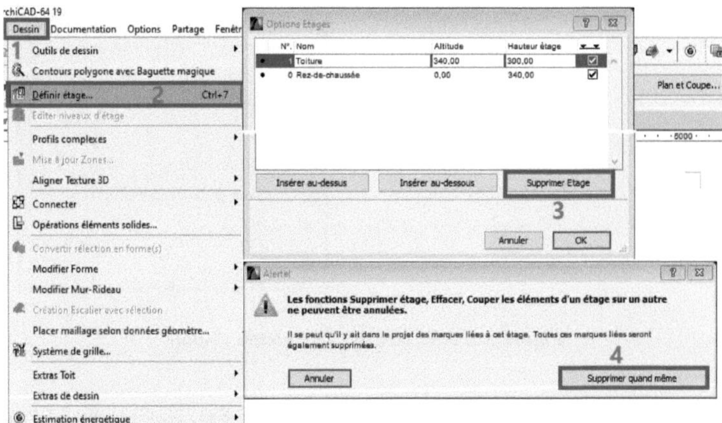


*Figure 4.0. Tips and procedures for selecting a single-level house*

Once the selection of the house level is complete, let's use the wall tool to start our very first drawing:

**Tips and case studies:**

☑ Select **wall** in toolbox ;

☑ Click on Default **wall options** ;

☑ Choose **the shape of your wall;**

☑ Choose **the thickness of** your **brick** wall;

☑ Choosing **the materials to** build your wall;

☑ Click **on the model** to choose the paint;

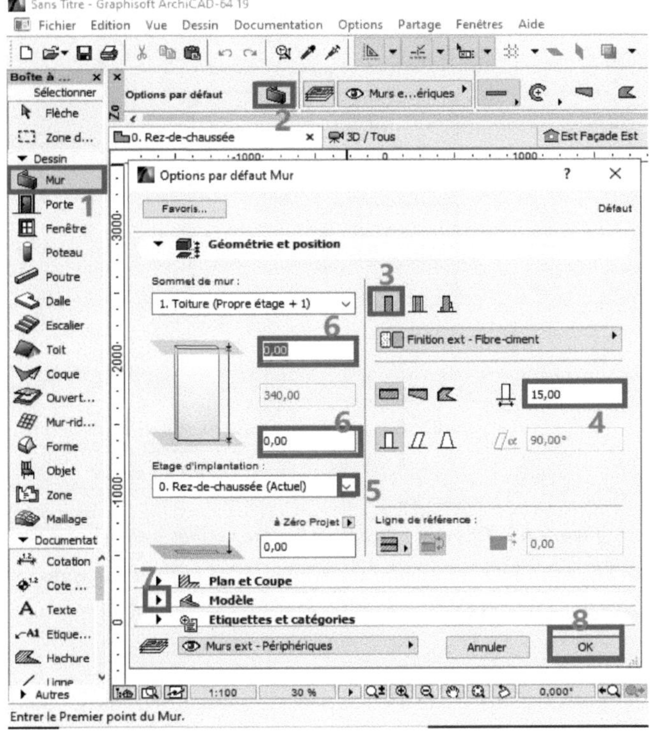

*Figure 4.1. Tips and procedures for using a wall*

☑ **Once the first wall has been set up, we can start drawing different walls repeatedly using these tools:**

**Remarks :**

☑ **Good to know**: once you've used a toolbox item, select **the** 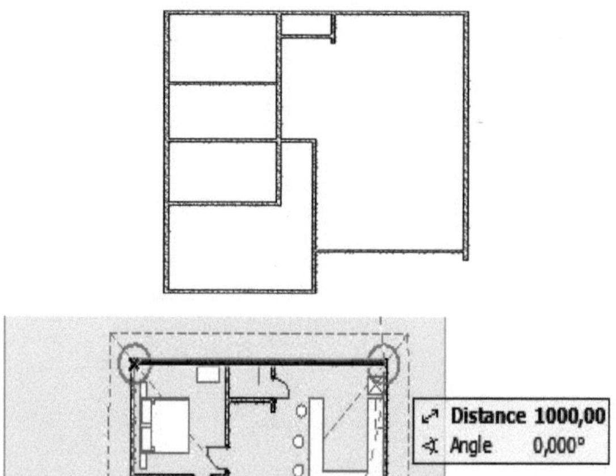 **arrow tool** for further work;

☑ Use **the Tab key** to enter the values for the **distances** and certain angles...of your various walls;

*Figure 4.2. Tips and procedures for the various walls shown*

☑ **Let's place the openings (doors and windows) on our various walls using the following practical tips:**

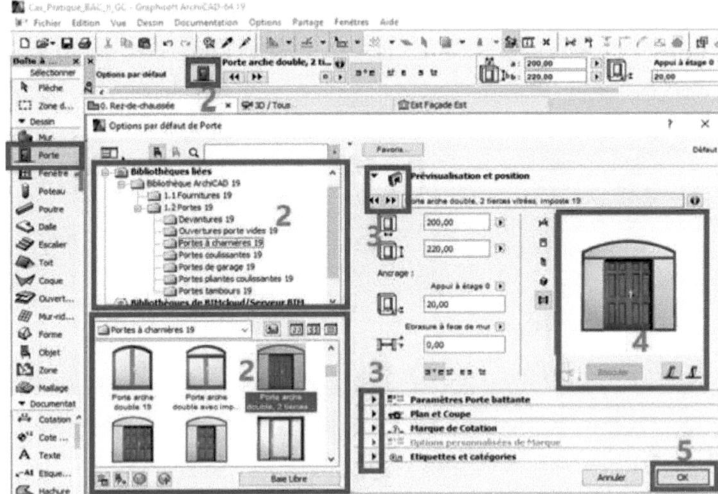

*Figure 4.3. Choice palette and door selection*

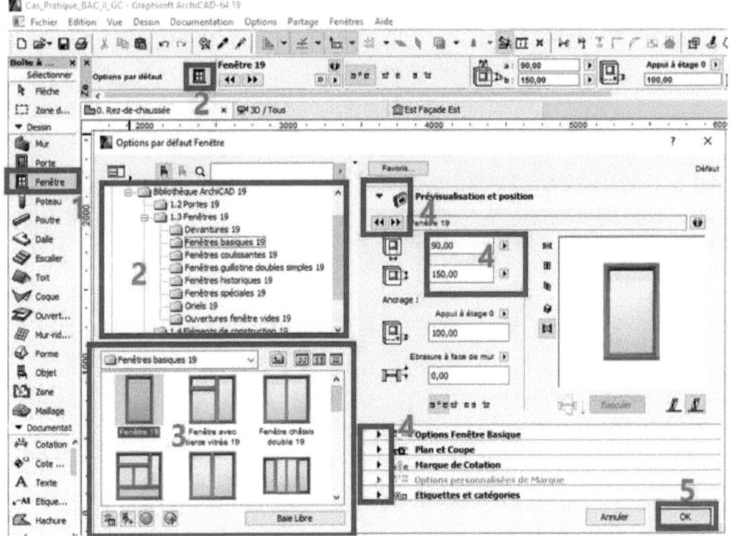

*Figure 4.4. Choice palette and Window selection*

☑ **Let's now place the slab** and visualize our **design in 2D** and **3D** before moving on to the layout;

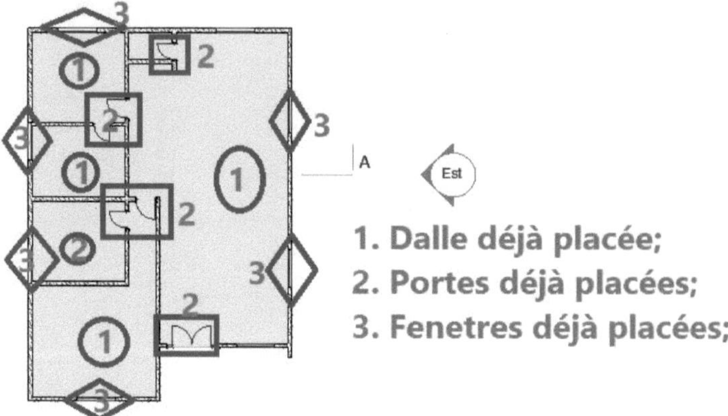



*Figure 4.5. Tips and procedures for the slab, doors and windows visualized in 2D*

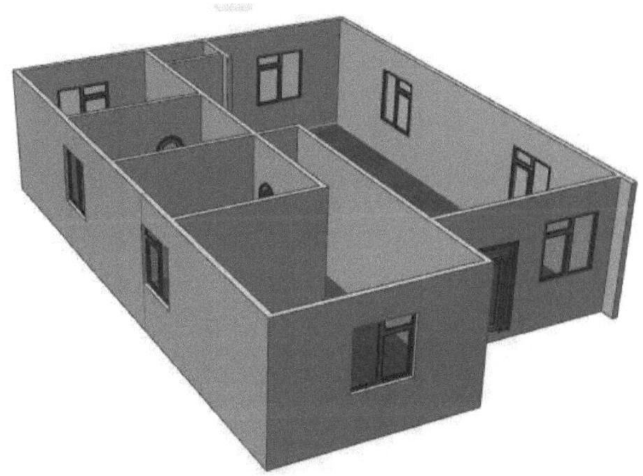

*Figure 4.6. Visualization of our first 3D drawing*

☑ **Let's furnish our Drawing with the following practical tips;**

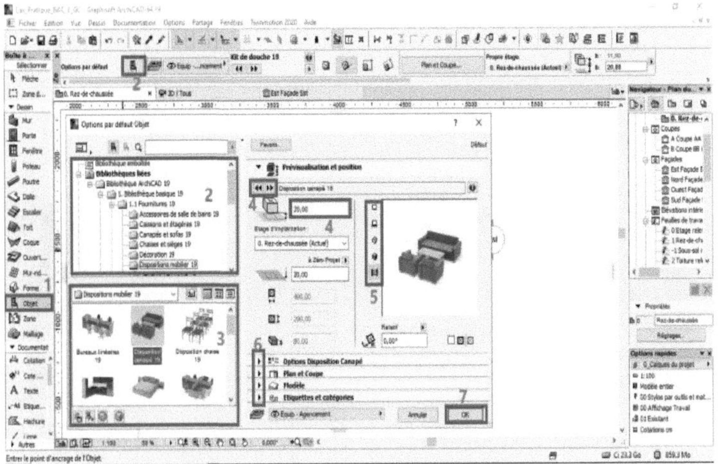

***Figure 4.7. Choice palette and selection of layout objects***

To fit out the house, use **the Object tool** in the **toolbox** on the left of the ArchiCAD environment, which contains various **libraries** as developed in **practical sheet no.03 point 13**.

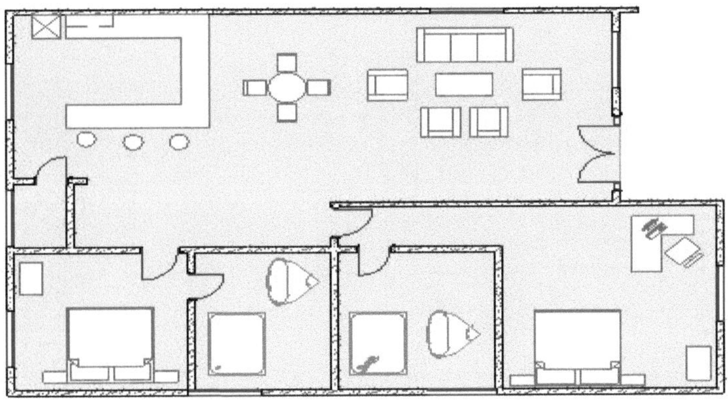

***Figure 4.8. View of our first 2D drawing after modifications***

81

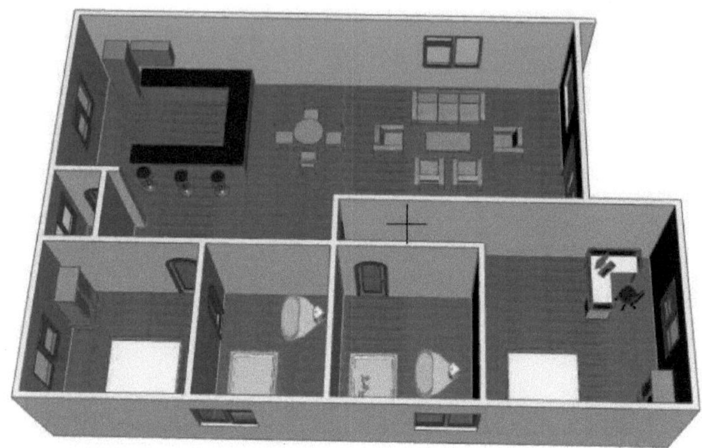

*Figure 4.9. View of our first 3D drawing after refurbishment*

☑ **Let's place the roof on our drawing;**

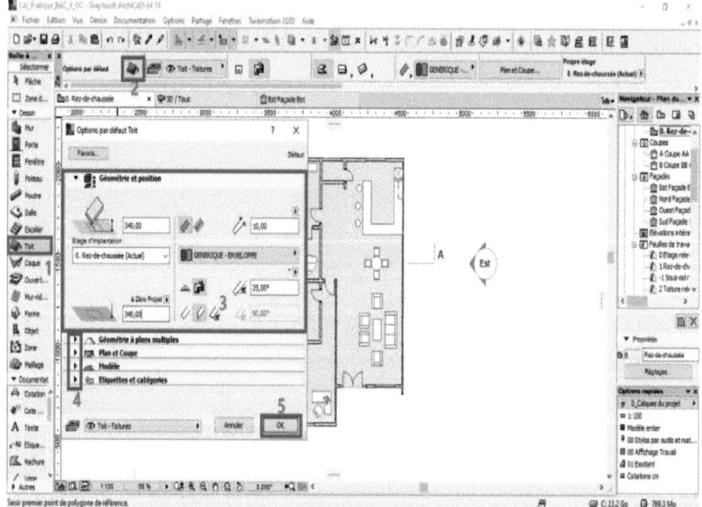

*Figure 4.10. Choice palette and roof selection*

**Good to know:**

☑ Roofing is installed all around the walls of your drawing;

☑ Use the tool ⬚ for **the construction method: Complex roof;**

☑ Use the ⬚ ▸ tool, with its drop-down list for **the group/rectangular gable construction method;**

☑ Use the ⬚ ▸ tool, with its drop-down list **for the construction method group / oriented rectangular gable or either for gable roofs;**

☑ Use the **F2 function** key once in 3D to return to 2D and **F3** once in 2D ;

☑ **Let's take a look at our first drawing in 2D and 3D after the roof has been installed;**

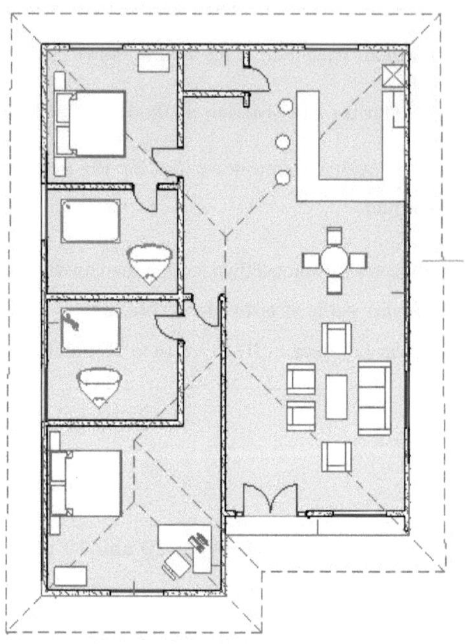

*Figure 4.11. View of our first 2D drawing after the roof has been installed*

*Figure 4.12. View of our first 3D drawing after the roof has been installed.*

☑ **Now let's take a look at the exterior of our first design;**

**Tips and case studies:**

- Selection of **wall tool** with settings such as (**Wall vertex** (Unbound), **40 cm altitude, 10 cm thickness**) ;

- In 2D, select the **Exterior face** Face extérieure , then start the exterior design;

- The first wall to be drawn, where the flowers will be placed, will have a **distance of 60 cm** per e.g ;

- Next, let's draw the **30 cm slab** inside the two walls we've drawn with a **"Peinture Vert-Foret" (Green-Foret** paint) model;

- Let's also draw **the outline of the project using the slab tool**;

- Let's place a seedling next to our drawing by clicking on the **object tool/Visualization/Environment/Garden/choosing a popular seedling** ;

☑ **Let's visualize our design after the exterior layout of our first design in 2D and 3D;**

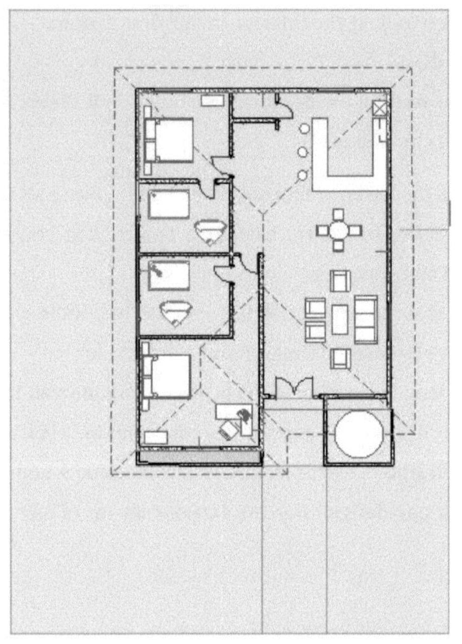

*Figure 4.12. View of our first 2D drawing after the exterior layout.*

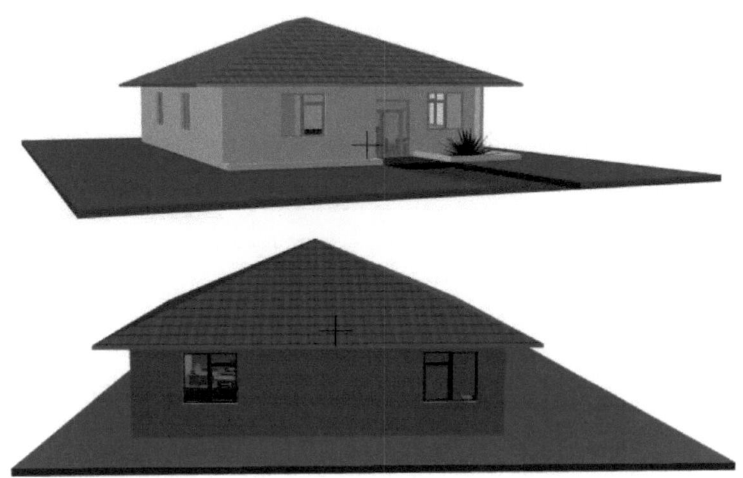

*Figure 4.13. Views of our first 3D drawing after the exterior fit-out*

**Good to know:**

☑ For photo rendering, let's continue with two other software programs, depending on the choice of the architect, using **Twinmotion** or **Lumion**;

☑ The user can save all the images in his design with other extensions such as **.jpg, .Pdf** etc. by clicking on **File/Save as** ;

**Note**: other difficulties can be encountered when installing **a gable roof,** for which we recommend the **following tips**:

Take, for example, the simple construction of a large room with the following elements:

- Building the walls of your room;
- Place the door and windows ;
- use the slab for paving;
- You can also design the exterior of your drawing;
- And a gable roof;

**Tips :**

The tips to be used for these tools in our second case study are identical to those developed in case study N°03.

☑ **Let's draw and visualize our design in 2D and 3D before installing the gable roof;**

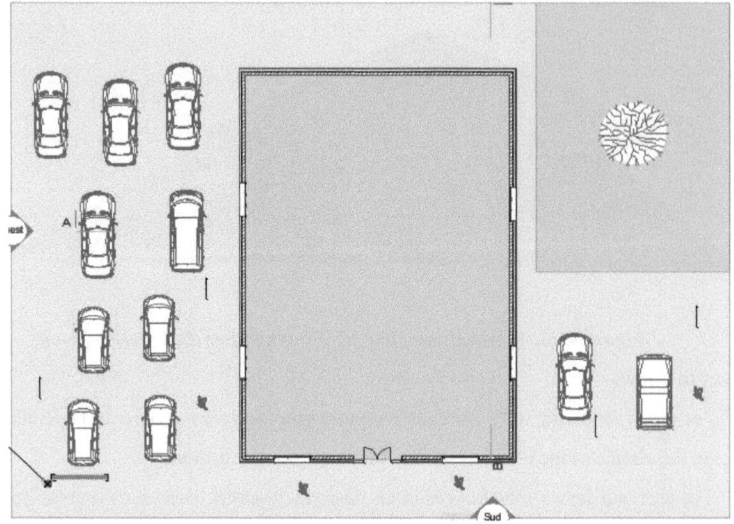

*Figure 4.14. 2D view of our great hall after drawing the walls, doors and windows, and even the exterior fittings.*

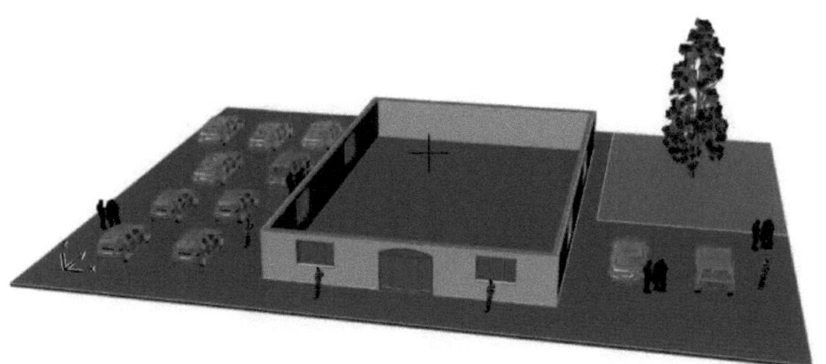

*Figure 4.15. 3D view of our great hall after drawing the walls, doors and windows, and even the exterior fittings.*

☑ **Let's place a gable roof on our drawing;**

**Tips :**

- The roofing tips for the design of our large room are the same as those used for the first case study;

- The peculiarity lies in the choice of the gable roof, where you must first click on the drawing tool , then in **the default option** select and move on to drawing ;

- Let's visualize our design in **2D and 3D** and see how it looks;

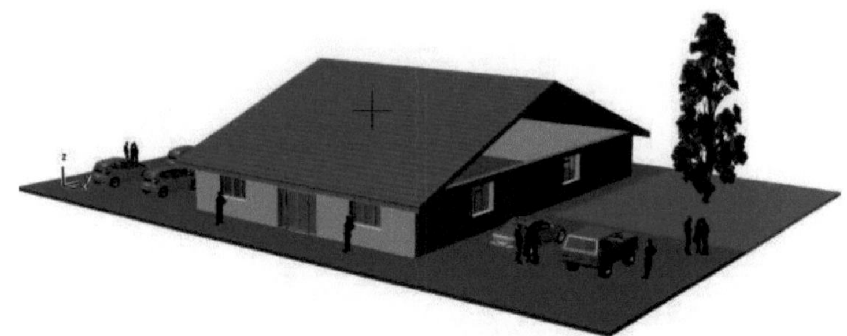

*Figure 4.16. 2D and 3D view of our large hall after the roof has been installed.*

After the roof was laid, there were gaps on both sides. Now we're going to connect the two walls to our roof and visualize how it's going to work;

**Tips :**

☑ Press **F3 to switch to 3D if you are in 2D**, select the **3D** walls where the voids are observed and move them to a height of more than **5 or 6m**;

☑ Click on **the wall drawing tool** and change the height of the walls;

☑ Re-select these walls, then **right-click** and choose the command **Connect/Solid operations**;

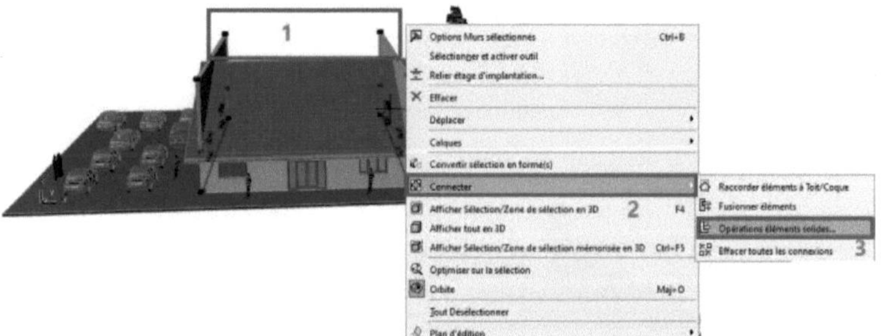


*Figure 4.17. Tips and procedures for connecting walls to the roof*

☑ In the dialog box that appears, the two selected walls appear as 2 targets;

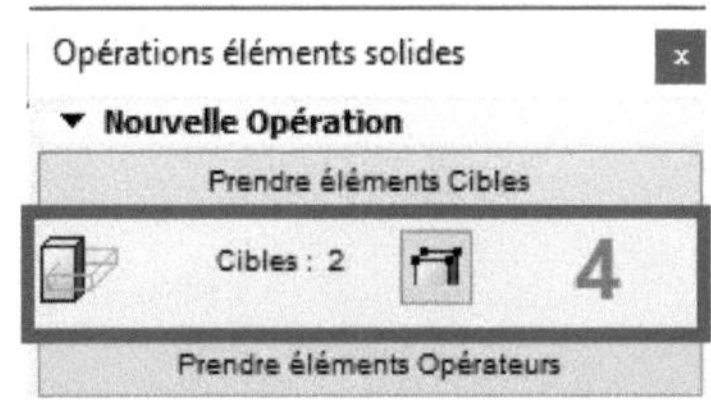

*Figure 4.18. Tips and procedures for selecting walls as target elements*

☑ After selecting the target walls, let's select the roof as the operator element;

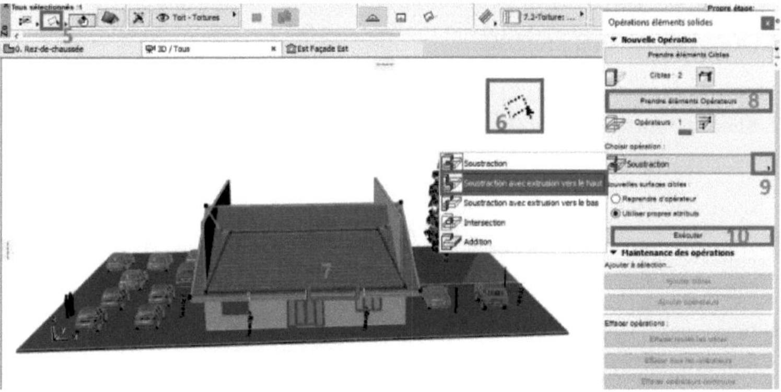

*Figure 4.19. Tips and procedures for selecting the roof as operator element*

☑ Finally, let's take a look at the final 3D drawing of our great hall after all the tricks have been applied;

***Figure 4.20. 2D and 3D views of our large hall after the roof has been installed***

**FLOOR DRAWING MODEL**

**Tips :**

☑ Start by setting the working units once again by going to **Options/Project Preferences/Working Units** and choosing **centimetres** or by using the keyboard shortcut **CTRL+F** ;

☑ After that, let's define the levels of our drawing, by doing Menu **Drawing/Define Floor ;**

☑ In the dialog box that appears, choose the different heights in the sense that 340m will be for the first floor, 320m for the other two levels and 300 for the roof; then start drawing;

☑ Once these settings are complete, draw the 4 walls of your building and place the parking for 2 vehicles and a staircase on the first floor;

☑ Use the keyboard shortcut **CTRL+T (Translation)** to **move** an object and CTR+R (for Rotation) ;

☑ Place **a 15 cm slab** on the pavement;

☑ Let's visualize **in 2D and 3D (F2 and F3)** the floor plan of our first floor;

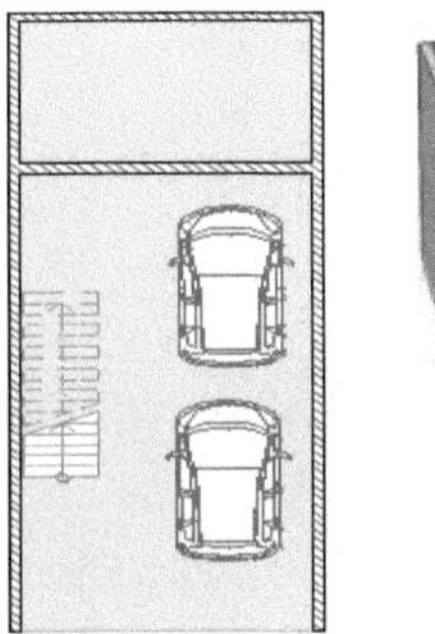 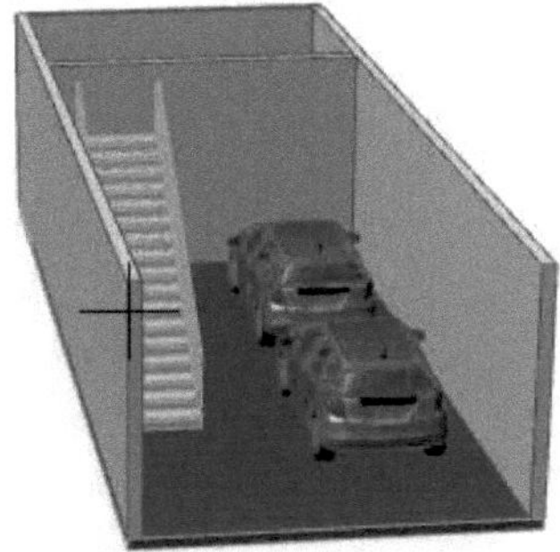

*Figure 4.21. 2D and 3D views of our floor plan on the first floor*

☑ To continue, let's move upstairs and see what happens;

☑ To do this: use **the tips below** to keep the same references as on the first floor;

- Go to **2D** and select **0.first floor** on the right of the screen and choose the **Show as track reference** option;

- These last tips will be identical when we move on to the design of the **second floor**;

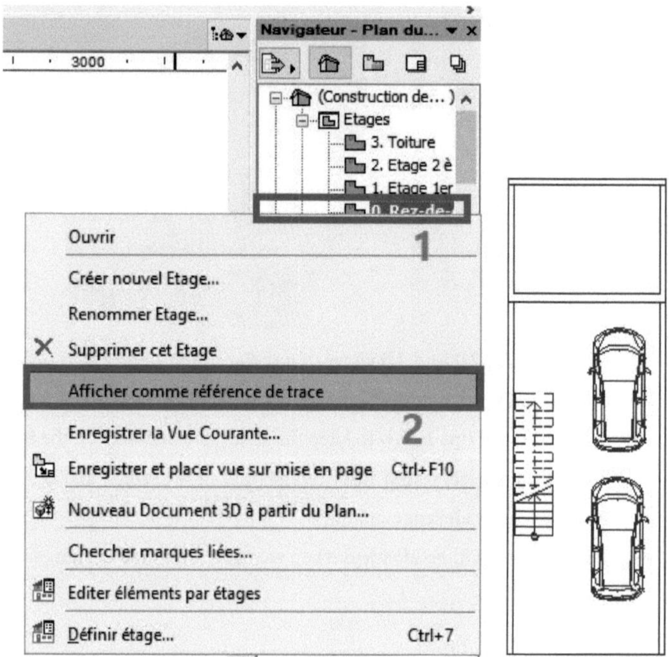

*Figure 4.22. Tips and Display for selecting the Show as trace reference option*

☑ Here's what you'll see after executing these tips;

☑ Let's place another **staircase on the** left-hand side and a 20 cm slab using the **geometry option** tool: **polygonal** for the second floor, then visualize the drawing **in 3D;**

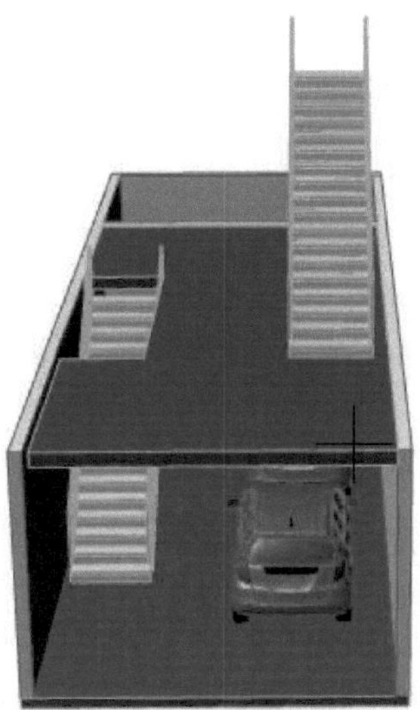

*Figure 4.23. 3D view of laid slab and staircase on second floor*

☑ Now let's make **the first floor background** disappear **at the 1ᵉʳ floor** level on the right by clicking on the **Show as source reference** option **and visualize in 2D and 3D**;

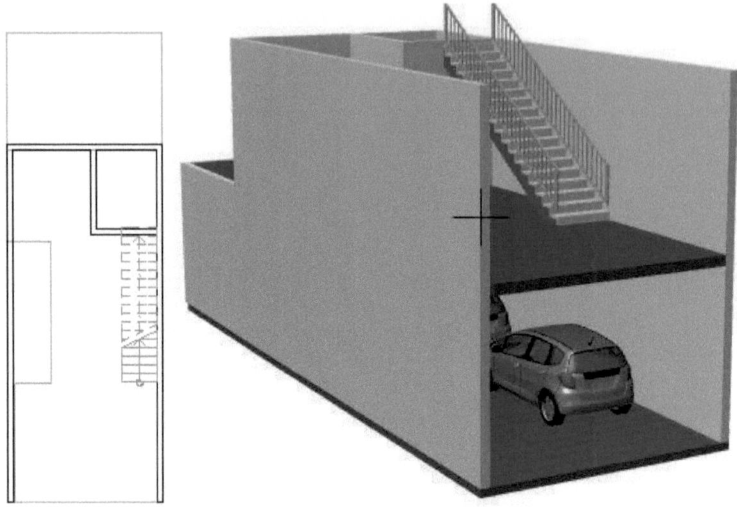

*Figure 4.24. 2D and 3D views of the second floor*

☑ Now let's place **a curtain wall and edit it**; but first let's use **a reference** in the **documentation** pane and choose the **line** tool ;

☑ To edit **the curtain wall**, simply go **to 3D (F3)**/select **the curtain wall** and an **Edit** command will appear on the selected **wall**;

☑ Click on the **default curtain wall option** and choose **layout/Primary lines/A** (Size 100) this is the width of the **curtain wall/Secondary lines/Sizes** (240, 80, 240, 60, 240) ;

☑ Click on Edit, and various commands appear (**OK, Cancel, Environment, Layout, Frame, Panel, Junction and Accessories**);

☑ In the information area, select (**Type: Generic/Mr 19 doors**) to change a panel to a balcony door or another type of your choice;

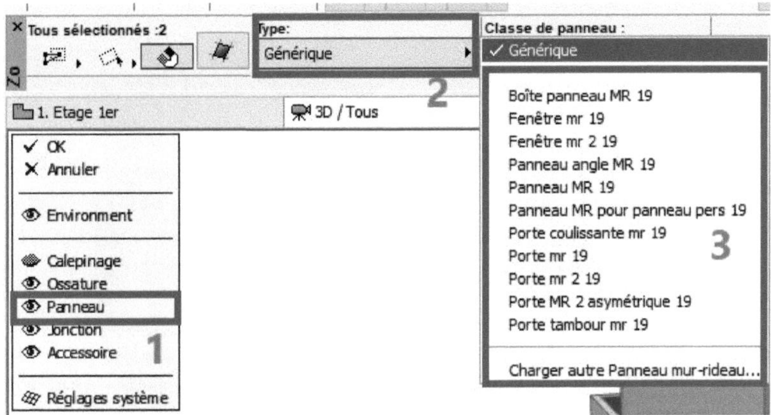


*Figure 4.25. Tips for turning a curtain wall panel into a door*

☑ Select a part of the slab to draw the balcony;

☑ Let's place a **railing (Barrière) on the balcony and staircase** of our second floor by clicking on : **Object tool/Choose building elements/Gates and railings** ;

☑ Let's move on to the second floor, with its **toilet (window), lounges (lower seat (without)) and kitchen (window);**

☑ Let's visualize **in 2D and 3D** and see what it will look like;

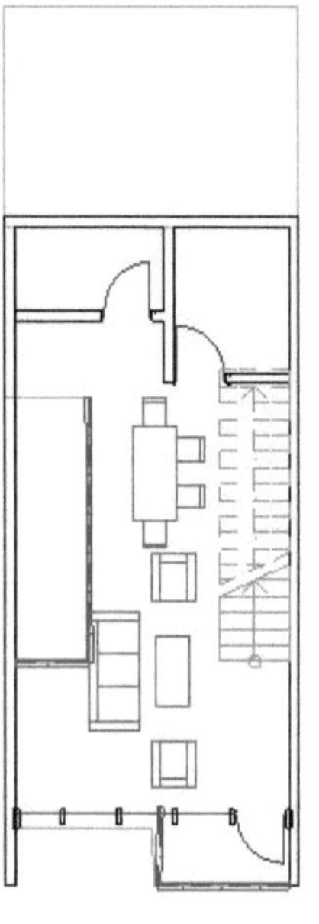 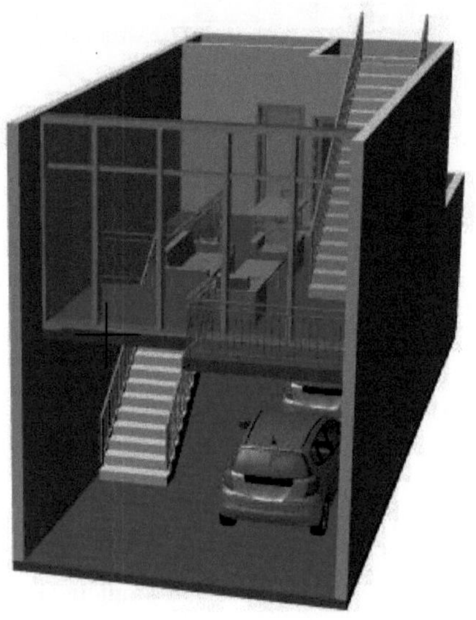

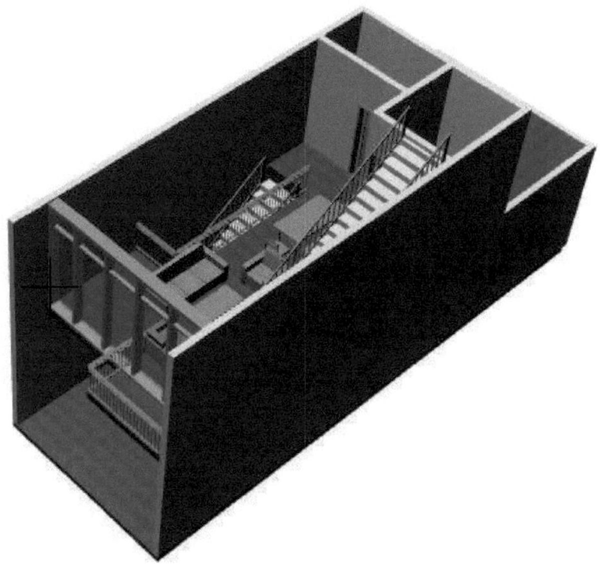

*Figure 4.26. 2D and 3D views of the first level*

☑ Let's move on to the drawing of the **second level of** our building;

☑ The procedures remain the same as for the second floor;

- Go to **2D by** pressing **F2** then select the **Second Floor** on the right;
- Then select **the second floor** and choose **Show as trace reference** then **Second floor** to start drawing;

☑ First, let's place **the slab** by clicking on **the slab tool, the walls and curtain walls, and the railing (Barrière) on the balcony and staircase**;

☑ Our **second floor** consists of **bedrooms, a study, two armchairs and a small table**;

☑ Let's visualize in **2D and then 3D** before installing the roof;

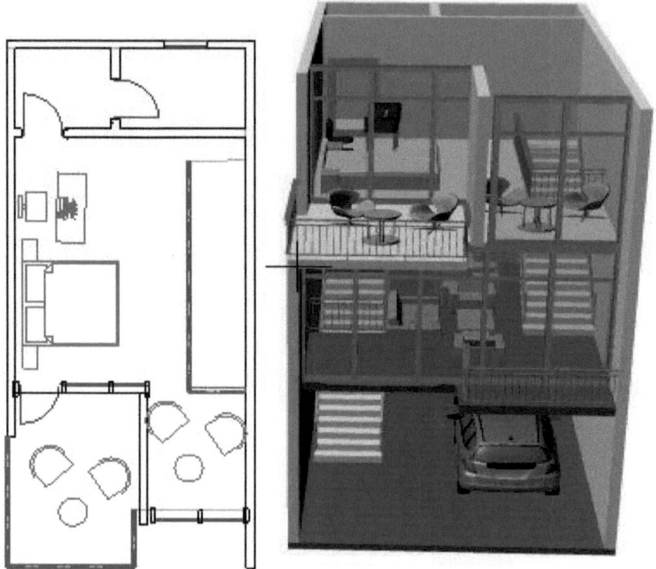

*Figure 4.27. 2D and 3D views of the second level*

☑ Finally, let's put a roof on it by selecting **Roofing** on the right (**Slab/Nature-white rendering**) and see what our design will look like;

☑ The procedures remain the same as for the second floor;

- Go **to 2D by** pressing **F2** then select **Roof** on the right;
- Then select the **Second Floor** and choose **Show as trace reference** then **Roof** to start drawing;

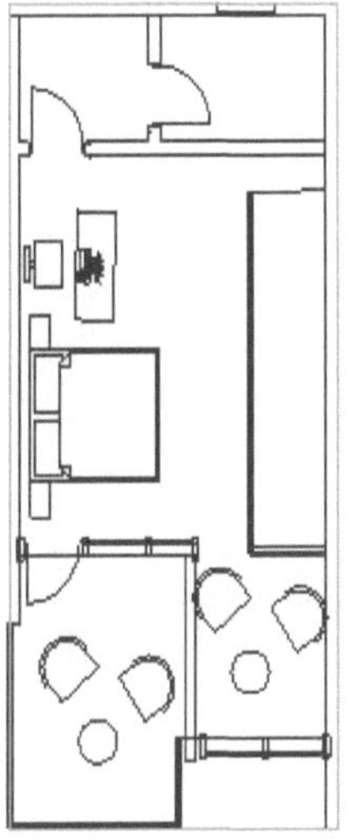 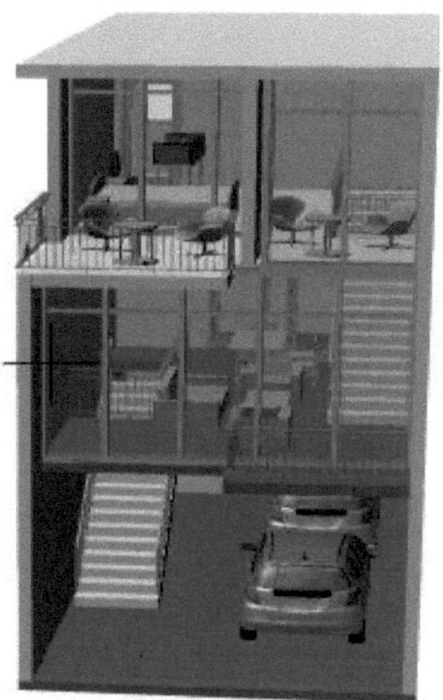

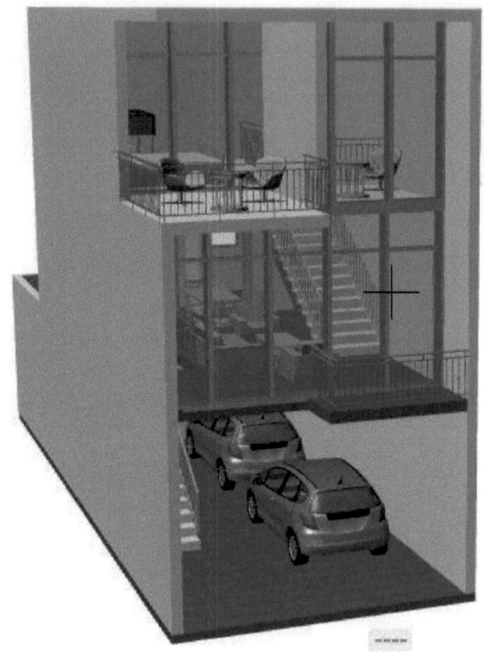

*Figure 4.28. 2D and 3D views of the second level after* roof *installation*

# CONCLUSION

On the face of it, the Practical Computer Aided Design: ArchiCAD course presents a modeling application that enables architects to design buildings productively through the Virtual Building concept. With ArchiCAD, architects can concentrate on their drawings, whether working individually or in collaboration with a team, thanks to the Teamwork function that facilitates data exchange with other professionals. This comprehensive and flexible BIM solution adopts the Building Information Modeling process for the creation and management of building data throughout its lifecycle.

What's more, mastering this software is crucial to the creation and management of the multiple models and drawings involved in an architectural project. Automated procedures are therefore essential to facilitate navigation between multiple documents and information, while keeping these elements updated and synchronized in real time. With ArchiCAD 19, efficient drawing management, instant access to all information and the ability to archive, share and distribute it are all assured. That said, the introduction of the BIM (Building Information Modeling) concept has marked a major turning point in the way architectural professionals approach the modeling and management of construction data. The ability to visualize projects holistically, simulate interactions and assimilate details through 3D and 2D representations represents a significant advance in the field of computer-aided design. ArchiCAD has succeeded in reconciling this modeling power with a user-friendly interface, offering a rewarding overall experience for architecture and construction professionals.

What's more, mastery of ArchiCAD opens up a vast field of possibilities in the way construction professionals manage the creation and management of architectural models. Whether designing, assembling, simulating interactions, quantifying or visualizing details, ArchiCAD proves to be a valuable and versatile tool that contributes to the efficiency and precision of the design process. As we conclude this hands-on exploration of ArchiCAD, we encourage students to continue to deepen their knowledge and explore the advanced applications of ArchiCAD in their professional practice. Stay curious, keep experimenting and apply the concepts and skills to enrich your future projects. Computer-aided design will continue to evolve, and your mastery of ArchiCAD positions you to be at the forefront of these developments.

All in all, ArchiCAD, through its 3D modeling and compatibility with Microsoft Windows and Mac OS operating systems, brings a major advance to the field of architectural design by offering 3D building simulation, known as BIM (Building Information Modeling). ArchiCAD's concepts enable efficient modeling, simulation, representation and quantification of project components, providing a global and detailed vision in plan, facade, section or 3D view.

# TUTORIAL

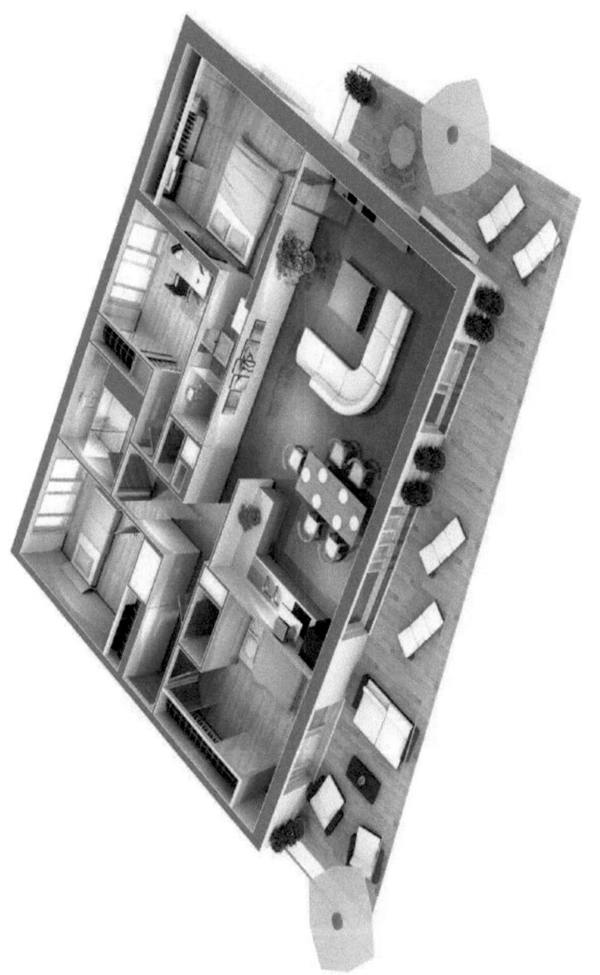

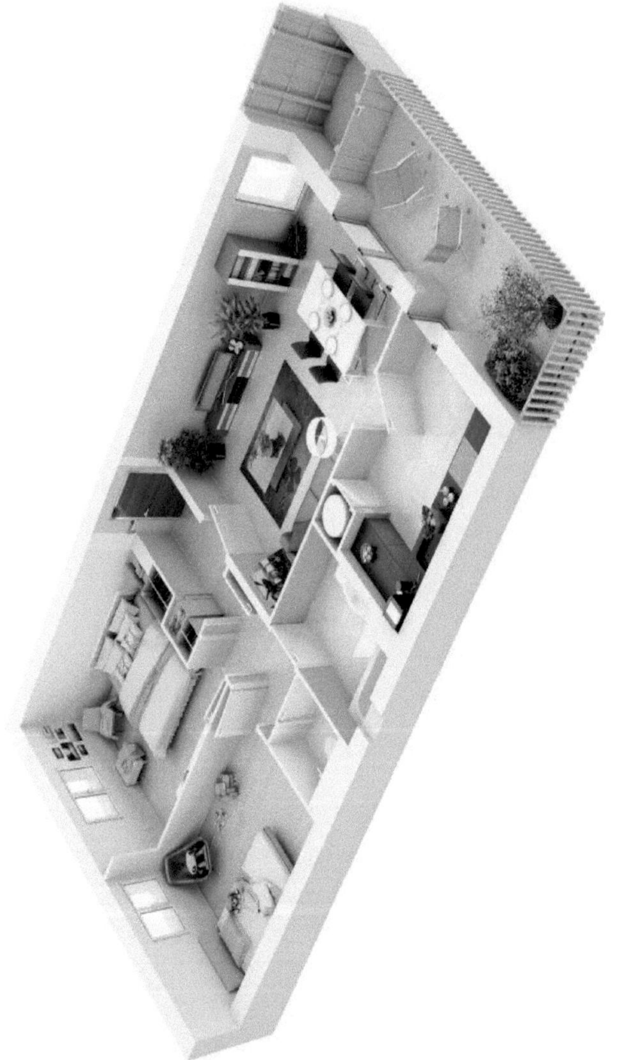

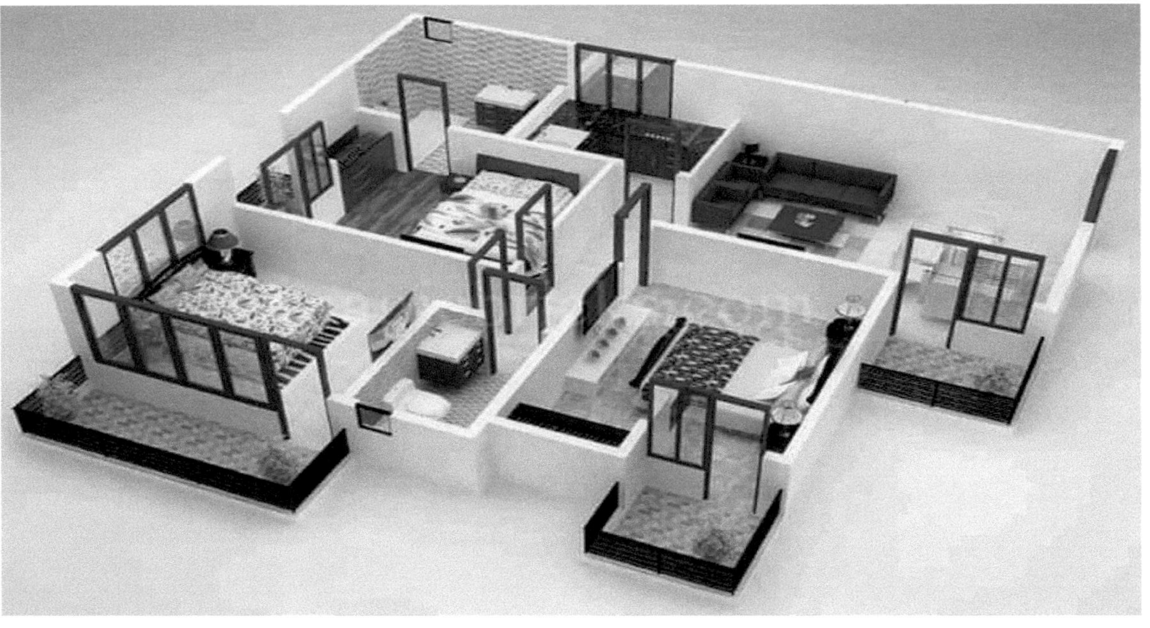

# TABLE OF CONTENTS

yes
**I want** morebooks!

Buy your books fast and straightforward online - at one of world's fastest growing online book stores! Environmentally sound due to Print-on-Demand technologies.

Buy your books online at
**www.morebooks.shop**

Kaufen Sie Ihre Bücher schnell und unkompliziert online – auf einer der am schnellsten wachsenden Buchhandelsplattformen weltweit! Dank Print-On-Demand umwelt- und ressourcenschonend produziert.

Bücher schneller online kaufen
**www.morebooks.shop**

Printed by Books on Demand GmbH, Norderstedt / Germany